ELEMENTS OF
COMBINATORIAL COMPUTING

To my parents and my MARvellous harem

ELEMENTS OF COMBINATORIAL COMPUTING

BY

MARK B. WELLS

Computer Science Research Group Leader, Los Alamos Scientific Laboratory
of the University California, Los Alamos, New Mexico 87544

PERGAMON PRESS

OXFORD · NEW YORK
TORONTO · SYDNEY · BRAUNSCHWEIG

Pergamon Press Ltd., Headington Hill Hall, Oxford
Pergamon Press Inc., Maxwell House, Fairview Park, Elmsford,
New York 10523
Pergamon of Canada Ltd., 207 Queen's Quay West, Toronto 1
Pergamon Press (Aust.) Pty. Ltd., 19a Boundary Street,
Rushcutters Bay, N.S.W. 2011, Australia
Vieweg & Sohn GmbH, Burgplatz 1, Braunschweig

First edition 1971
Library of Congress Catalog Card No. 77–129633

Printed in Hungary

08 016091 3

CONTENTS

PREFACE

THE development of high-speed computing over the past two decades has had an extraordinary impact upon the scientific community. At first, this "computer revolution" most strongly influenced, and was influenced by, physicists, engineers, numerical analysts, and other applied scientists interested in performing arithmetic calculations to obtain numerical answers to their problems. However, the versatility of the electronic computer, its ability (crude and unplanned though it was and generally still is) to perform nonarithmetic manipulations, soon stimulated interest among "pure" mathematicians—primarily, number theorists and algebraists—and among scientists of less numerically oriented disciplines such as operations research and linguistics. At present, "nonnumerical computing"—computing in which the basic operations are nonarithmetic (e.g. logical, set-theoretic, symbol manipulative) and in which classical numerical analysis plays, if at all, a minor role—is gaining in importance. This book is about *one* aspect of nonnumerical computing: the high-speed manipulation (e.g. generation, covering, comparison) of combinatorial objects such as permutations, partitions, and linear graphs.

There is, of course, a close kinship between this *combinatorial computing* and combinatorial mathematics. In fact, most of the methods and examples appearing in this book are taken from the study of problems in combinatorial theory. Nevertheless, there is important interaction between such computing and other disciplines. The techniques presented here find application in the fields of number theory, algebra, switching theory, biomathematics, game theory, and operations research, as well as combinatorics.

The primary purpose of this book is to bring together under one cover the basic and important computer methods of solving problems which involve the handling of combinatorial entities (more generally, of finite sets). Moreover, we wish to assist the reader in gaining facility for constructing combinatorial algorithms of his own design. Thus special attention is given to the structure of the algorithm, why they work, and how they may be altered to accomplish similar but distinct tasks. Of course, we also hope that this book will be a source of ideas for workers in combinatorics *and* computer science. Two underlying goals of this book are to introduce the pure mathematician to the possibilities of high-speed computation, and to point out to the computer scientist some of the needs of the research mathematician.

With regard to these objectives, the language used for describing the algorithms is of central importance; and there are several considerations in choosing an algorithmic language. Firstly, the language should be natural, i.e., it should be readable and reasonably concise, using established mathematical notation as much as possible. Use of an uncontrived, not altogether foreign, language simplifies and enhances understanding of the algorithms and eases the task of their modification. Secondly, the language should have a degree of sophistication which allows suppression of detail inessential to conceptual understanding of the processes. Certainly a user should not be required to repeat in detail the thought processes previously followed by other qualified algorithmists or computer designers. Finally, it is desirable to have the algorithmic language serve as the programming language, i.e., to have facilities available for the automatic translation of the algorithms into "absolute code" for a particular computer. This, of course, greatly shortens the time from problem conception to the production of useful results. (This capability should play a subordinate role in the *design* of the algorithmic language, however.)

Algorithms are presented in this book in a language especially designed for combinatorial computing. This language is both more natural and more sophisticated than the common programming languages Fortran and Algol. Also, most of its features have indeed been implemented—as part of the programming language Madcap for the Maniac II computer in Los Alamos. The language and its implementation are discussed in Chapters 1 and 2. Chapter 3 is both an introduction to combinatorial terminology and to data representation within a computer. The serious discussion of algorithms and their structure begins with Chapter 4. An attempt has been made to use notations from the language in textual descriptions of the algorithms as well as in the programs which implement the algorithms.

Translation from the notation of this book to a programming language available on a specific computer should cause little difficulty, particularly if certain basic nonnumerical operations such as logical *or*, logical *and*, shifting, and counting 1-bits are readily programmable for that machine. To simplify this translation, care has been taken to define sophisticated notations in terms of previously defined, more elementary notations which closely resemble familiar Fortran and Algol statements.

Actually, sophistication is a relative matter. The science of computing and its language of presentation, like mathematics, grows in steps, the methods and notation at one level being the building blocks for higher order methods and notation. With the techniques and language presented here, we attempt to take but the first step in the founding of a science of combinatorial computing—the use of the word "elements" in the title of this book is certainly appropriate. We hope, however, to have achieved a unification of many of the basic concepts which can easily lead to more advanced methods and notations for combinatorial computing.

This book is intended primarily for mathematically trained individuals interested in the use of high-speed digital computers for obtaining "answers" (counter-examples and heuristic data perhaps more so than final results) to combinatorial problems. Beyond basic mathematical training—finite set theory and the elements of algebra and number theory (number systems, groups, congruences, etc.)—this book presupposes only a grasp of the concept of algorithmic computing. Actually, much computer and combinatorial terminology with which a person with this training could be expected to be familiar is herein explained anew. Just as the numerical analyst finds it necessary to employ difference equations when seeking help from a digital computer, the combinatorial analyst also finds it necessary to return to fundamentals for preparation of his computer programs. The fundamentals here, with which it is assumed the reader *has* facility, are the principle of induction (more precisely, the concept of recursive definition) and the logical *rule-of-sum* and *rule-of-product* (in a set-theoretic language, these are $A \cap B = 0$ implies $|A \cup B| = |A| + |B|$ and $|A \times B| = |A| \cdot |B|$, respectively). However, let the reader be warned that the elementary nature of the tools does not necessarily imply simplicity of the structures built with them. Many combinatorial algorithms are extremely complicated, and their mastery involves much plain hard work.

Discussion of the algorithms is for the most part casual; formal proofs are not included. Nevertheless, this book is appropriate as a supplementary textbook for an upper class or graduate course in combinatorial theory, computing, or numerical analysis. (In this respect it should be noted that the order of Chapters 4, 5, and 6 is somewhat arbitrary, as is the order of the major sections in Chapters 6 and 7.) Of course, this book could also be used as a *basis* for a course entitled, say, Introduction to Discrete Computing; although, for such an endeavor to be truly successful, an implementation of the language should be available to the student.[†] Indeed, it is our hope that many of the language features used here will some day be incorporated into a generally available programming language.

The exercises at the end of each major section are considered an integral part of the text. They should be read, even though not solved, since besides practice problems they contain definition of terminology, discussion of alternate methods and useful modifications, and suggestions for further research. Some of the programming exercises anticipate later discussion; hence their answers appear in the text. Most problems, however, are best pursued by actual computer experience. Answers to a few of these problems appear in Appendix II as tables calculated by the author. Asterisks are used to mark more difficult exercises, those which might better be called *projects*.

The single bibliography at the end of the book is organized by chapter. It contains suggestions for further reading as well as cited references. Many of the works are listed by virtue of their expository character and the extensive bibliographies which they contain.

All of the procedures presented in this book have been tested on the Maniac II com-

---

[†] The author, as a Visiting Professor, used the manuscript of this book for a senior level course under the Department of Mathematical Sciences at Rice University, Houston, Texas, during the spring semester 1970. Translation to Burroughs Compatible Algol for the B5500 computer at Rice was annoying but not difficult.

There is somewhat more material in this text than can be covered in one semester. A full year's course entitled "Combinatorics and Computing" could be fashioned from this book by introducing certain concepts of theoretical combinatorial analysis (e.g. generating functions, inclusion–exclusion, Polya enumeration) at appropriate points in the presentation.

puter at the University of California Los Alamos Scientific Laboratory, Los Alamos, New Mexico. However, since there is still the possibility of inaccuracies due to debugging oversights, language ambiguities, or other errors, the reader is advised to take nothing for granted.

Many people have assisted me in diverse ways in the preparation of this work. First, I would like to thank the administration of the Los Alamos Scientific Laboratory for allowing me to undertake this project under their sponsorship. Discussions with Stanislaw Ulam, Roger Lazarus, Paul Stein, Robert Bivins, Nicholas Metropolis, Myron Stein, Marvin Wunderlich, Robert Korfhage, Donald Knuth, and Robert Floyd influenced this work. Particular thanks are due to Robert Bivins, Roger Lazarus, Nicholas Metropolis, and Paul Stein for their constructive comments upon reading the manuscript. I am indebted to Verna Gardiner for debugging assistance and to Donald Bradford for his patient help with the language development as well as for debugging assistance. The final manuscript was expertly typed by Margery McCormick and Dorothy Camillo. I also wish to thank Fred Cornwell, Jay London, and Jane Rasmussen for their clerical support. Finally, along a somewhat different vein, I would like to acknowledge two especial debts—to Professor S. Ulam for his continued stimulation of my work and to Professor D. H. Lehmer whose commonsense approach to mathematics and computing has always been an inspiration to me.

This work was performed under the auspices of the United States Atomic Energy Commission.

CHAPTER 1

A LANGUAGE FOR COMBINATORIAL COMPUTING

THE vocabulary of present-day computers is such that algorithms written in language intelligible to most scientists *cannot* be presented directly to the computer for execution —a translation is required. However, auxiliary computer programs can be written to accomplish "automatically" at least part of this translation. Such programs are called *compilers*; they translate a *source program*, written in *source language* (programming language), to an *absolute machine code* directly executable by the computer. It is desirable that the algorithmic language of the scientist and the source language be nearly identical, for then no translation by erring humans is required.

The profit from a sophisticated programming language in which concepts may be expressed directly is considerable. The programmer who wishes to indicate an "iteration" (see § 1.4) should no more have to be concerned with the details of setting, incrementing, and testing the dummy index than the mathematician should have to write

$$\lim_{h \to 0} \frac{f(x+h)-f(x)}{h}$$

when he wishes to indicate the derivative of a function with respect to x. On the other hand, even if it were possible, it is probably not reasonable (due to efficiency considerations) to allow programming language development to proceed without regard to automatic translation of the language to computer code. In this book a language is used which is more sophisticated (especially for combinatorial computing) than the common programming languages Fortran and Algol, yet which also lends itself to efficient translation by a compiler.

This language is described, quite informally, in this chapter. It is to be hoped that a mathematically trained individual having some familiarity with programming languages in general could skim rather lightly over this material. Efficient programming and certain aspects of the translation problem itself are discussed in Chapter 2.

1.1. Fundamentals

Our combinatorial language is a *statement language*, i.e. a (source) program for performing a certain task consists of a sequence of statements describing the calculation which is to take place. It is "user-oriented" in that standard mathematical notation and English, not foreign to an uninitiated user, are the basis of statement construction.

1.1.1. *Program Structure*

The statements encountered in a program are of two types: *formulae* and *control clauses*. Most requisite computation is specified within a formula, while control clauses (conditional statements and iteration quantifiers—see later sections) govern the logical flow of the program. Formulae are usually written one per line, but may appear several per line separated by semicolons (semicolons are generally omitted from the end of a line). To indicate continuation onto the next line, very long statements are broken immediately following an operation symbol, the "equals" symbol (used as the assignment operator—see below), or a comma. The statements under jurisdiction of a control clause are separated from that clause by a colon. Such statements which are written on separate lines are indented in order to display the extent of the jurisdiction. Indentation within indentation often occurs to several levels. The case where the number of levels is variable or indefinite is discussed in § 1.5.

For *internal* reference purposes (e.g. reference by transfer of control statements —see § 1.3), any statement may be given a label consisting of an unsigned integer preceded by the number symbol #. These labels are placed in a special column just to the left of the statement sequence.

An illustration of the structure of a program is given in Fig. 1.1. (Note: Fully capitalized words such as FORMULA, STATEMENTS, PROGRAM, ARGUMENTS are used in place of specific examples when program structure rather than notational detail is being emphasized.)

A program often consists of several distinct *segments*, each segment designed to accomplish a particular part of the total job. For delimitation and reference purposes (e.g. describing program changes), segments are labeled with capital letters A, B, ..., and lines of program within a segment are then numbered 0, 1, ...; a decimal point separates segment labels from line numbers. When exhibited, these segment and line labels appear at the far left as illustrated in Fig. 1.1.

A segment may be used as a "procedure" (a finished program referenced via functional notation—see § 1.1.4), in which case line 0 consists of the procedure name and list of arguments. Subsegments, labeled for example A.A., C.B., A.B.A., though useful in general, appear infrequently in this book. The nesting of segments and the "local–global" properties of names and statement labels used within a segment are discussed in § 2.4.

A.1   FORMULA; FORMULA

A.2 #1   CONTROL CLAUSE:

A.3   FORMULAE

A.4   CONTROL CLAUSE: FORMULA

A.5 #2   STATEMENT

A.6   STATEMENT

.

.   PROGRAM

.

B.0 PROCEDURE NAME(ARGUMENTS)

B.1   STATEMENT

FIG. 1.1. Sample program structure.

A formula takes the form
$$v = E \qquad (\text{or } E \rightarrow v),$$

where v represents the name of a variable which is to be assigned the "numerical value" as computed from the mathematical expression E. (The use of the equals symbol as an assignment operator is a slight aberration from normal mathematical usage due to the admission of formulae such as $i = i+1$. We avoid this construction by use of von Neumann's $i+1 \rightarrow i$. However, in general, this use of the equals symbol should cause no confusion. Symbols used by other languages for this assignment operation include: $:=$, $\leftarrow$, and let ... $=$.) For names of variables, we use (1) the English letters in upper and lower case, A, B, ..., Z, a, b, ..., z, (2) these letters modified by a bar, tilde, star (asterisk), and/or an Arabic numeral, e.g. $\bar{r}$, $\tilde{A}$, x^*, $p3^*$, and $H2$, and (3) the upper case English letters modified by a few lower case letters, e.g. Tag, $Link$, and Gt. These naming conventions give us ample freedom in choosing variable names without exceeding the character limit of a normal mathematical typewriter. Also, with only a little care (which we indeed exercise), juxtaposition may be used to indicate multiplication.

A statement of the form
$$v \leftrightarrow u$$

(where v and u represent names of variables) is equivalent to the following sequence of three formulae:
$$d = u; \quad u = v; \quad v = d$$

with d used here as a dummy variable name. A statement of the form
$$x,y,z = E$$

is equivalent to

$$x = E; \quad y = E; \quad z = E,$$

where E may be any mathematical expression (of course, a good compiler would produce machine code computing E only once). By way of terminology a formula

$$u \rightarrow v$$

is sometimes called a *transmission* (of the value of u to the "storage location" assigned to the variable v).

For reasons of efficiency we introduce the special "two result" assignment formula

$$\text{define } v,u \text{ by } E1/E2$$

into the language. For positive integral-valued expressions $E1$ and $E2$, this is equivalent to

$$v = [E1/E2]; \quad u = E1 - vE2,$$

where the square brackets denote the largest integer not exceeding the quotient.

Comments are usually given within the running text of the book proper rather than as part of the programs themselves. However, when, for clarity, commentary is included within a program, it is distinguished from executable statements by being enclosed in quotation marks and written well removed from these statements:

STATEMENT   "H is the complement graph"

PROGRAM

"The following computes T^*"

The assignment of all storage is considered to be the responsibility of the compiler or supervisory system, hence no storage-allocation statements are included in the language (see § 2.1.3).

1.1.2. Notation for Real Number Operations

The basic data on which digital computers operate are real numbers, more precisely, rational approximations to real numbers, including the integers as a subset. (Of course, the precision of the real number approximations and the maximum magnitude of the integers depends on the word length and word organization of a particular computer.) The standard arithmetic operations on these quantities—addition, subtraction, multiplication, division, exponentiation, and root extraction—are indicated in the language in the usual mathematical notation. Multiplication is implied by the juxtaposition of factors as well as by explicit use of the cross symbol. Parentheses and square brackets are used for grouping.

In certain instances, however, namely when the bracketed quantity is a single term (e.g. $[x]$, $[a/b]$, $[(x-1/2)]$), square brackets are used to denote the number-theoretic function "greatest integer in". In these cases, the customary definition (e.g. $[3.1] \equiv 3$, $[-3.1] \equiv -4$) obtains. (While satisfactory for the combinatorial language as used in this book, this dual use of square brackets is not acceptable for a general purpose programming language as serious ambiguity can arise in complicated arithmetic expressions. Other languages use special function mnemonics or notation such as $entier(x)$, $ipt(x)$, $floor(x)$, or $\lfloor x \rfloor$ for this important function. Perhaps the best solution is to allow the programmer himself to declare or redefine special notations within a given segment of program—e.g. square brackets could imply ordinary grouping unless otherwise stated.) Other common functional notations employed are the bar notation, e.g. $|a|$, for "absolute value of", the exclamation point notation, e.g. $n!$ or $(a+b)!$, for "factorial of", and the display notation, e.g. $\binom{n}{r}$, for the binomial coefficients.

Another notation borrowed from number theory is that for modular arithmetic. For $n > 0$ the occurrence of the phrase (mod n) at a particular parenthesis level says that the numerical result for that level is to be replaced by its least positive residue modulo n. Thus if R represents that result, it is replaced by r, where $r = R - [R/n]n$. For example, when $n = 3$ and $a = 5$, the expression

$$a^2 - 2 \;(\text{mod } n)$$

has the value 2, while the expression

$$\left(a^2 \;(\text{mod } n)\right) - 2$$

has the value -1. This notation is most often applied when all quantities involved are integral, although it is not restricted to this case. The notation is undefined for a non-positive modulus.

The language includes a mnemonic functional notation for expressing common functions of real variables. In this book we use $log(x)$ for the natural logarithm of x, $log_2(x)$ for the logarithm to the base 2 of x, $exp(y)$ for the exponential function (i.e. $exp(log(x)) = x$), and $max(a, b)$ (and $min(a, b)$) for the maximum (and minimum) of the numbers a and b. The no-argument function $random$ selects a number x arbitrarily from the interval $0 < x < 1$—see § 2.2.4. These functions are "library functions", i.e. they are preprogrammed procedures existing as part of the compiling system (§ 2.4) which are automatically inserted into any program which uses them. Less common functions of real numbers and functions of other data-types (e.g. sets) using functional notation are introduced at various places throughout the book, see § 1.2 in particular. Notation for finding the maximum element of an array of numbers is discussed in § 1.4.5.

It happens frequently in combinatorial computing that more than one computer word is required to represent the monstrous integers obtained during a calculation. If standard real number operations only are to be applied to these quantities, then the notation used to express these operations remains unchanged (see § 2.1.2). However, if "multiple precision" is used to construct special representations on which peculiar operations are required, then the notation depends on the particular case at hand —Chapter 3 pursues this subject. When the time comes to program an algorithm for a particular computer, the reader should have little difficulty determining which variables require a multiple-word form.

1.1.3. *Data-types and Arrays*

In presenting algorithms for digital computer digestion, it is convenient to work with various forms of data besides real numbers—complex numbers, matrices, character strings, and (especially for combinatorial computing) sets, graphs, etc. In this book we use three basic *data-types*: real numbers (including the integers as a subset), unordered sets of natural numbers, and arrays whose entries are either real numbers or sets of natural numbers (a particular array has entries of only one type). (The elementary algorithms based on the data-types presented in this book will certainly suggest to the reader higher level data-types and thus simplified notation for the given manipulations. For instance, a linear algebraic "matrix" data-type certainly belongs in a language for combinatorial computing of a more advanced nature than is presented in this book. Sophistications of this sort will quite likely play an important role in future programming language evolution.) The variable naming conventions mentioned in § 1.1.1 apply regardless of the data-type of the variable being named. The data-type associated with a particular name remains constant during the presentation of a particular program but may change for other discussions (see § 2.4).

The referencing of an entry of an *n*-dimensional array (an *n-array*) requires the specification of *n* integral coordinates. These coordinates are placed as subscripts, separated by commas, on the name of the array; for example, $A_{i,j}$ is the entry in the *i*th "row" and *j*th "column" of the 2-array named *A*. This quantity may be a real number or a set of natural numbers. Arrays occurring in combinatorial computing are frequently nonrectangular, in fact the range of coordinate values in one dimension may depend on the coordinate values in other dimensions. Declarations of coordinate range and of entry data-type are generally not required as part of a source program (see § 2.1).

Constants of real type are written in customary decimal notation: 6.2, 0, −17, 65536, 0.1469. Unordered set constants are primarily (but see also § 1.2.1) written in the common tabulated notation: {0,6,50}, {25}, { }; while ordered set constants, constant 1-arrays, are written, when $n \geqslant 2$, in *n*-tuple notation: (5,4,3,8), (7.1,4.3,0.2), ({0}, {6,10}, { }, {99,100}). The coordinate range of an *n*-tuple is considered to be 0 to $n-1$.

Although we use the standard terminology "vector" and "matrix" (for 1-array and rectangular 2-array respectively), no linear algebraic definitions of the arithmetic operations upon arrays are implied. Operations upon arrays are generally expressed by iteration of formulae involving individual entries of the arrays. However, we sometimes omit the iteration specification (§ 1.4) and the subscripted indices when the same operation applies to each entry of an array or to corresponding individual entries of several operand arrays. For example, if A, B, and C are arrays with the same subscript (coordinate) ranges, the statement

$$A = B+C$$

would set each entry of A equal to the sum of corresponding entries of the B and C arrays.

Of course, this shorthand notation can only be used when the subscript ranges are known. Thus, statements such as

$$B_{1 \text{ to } n} = 0 \tag{1.1}$$

and

$$D_{0 \text{ to } 3} = (\{ \ \}, \{0\}, \{1\}, \{0,13\}) \tag{1.2}$$

are sometimes used to specify the subscript range for an array *and* to initialize (or fix) its entries. (In the first example, B_1, B_2, ..., B_n would each be set to zero.) The appearance of a restricted range specification at a subscript position of an array with an established range indicates that only part of the array is being referenced. For example, if $A_{0 \text{ to } 99, 0 \text{ to } 99}$ is a known 2-array, then $A_{0 \text{ to } 49, j}$ is the first half of the jth column of the array A; it is itself a 1-array. If an entire subscript range is to be used, a dash may be written in place of a 1 that range specification. For example, $A_{i, \_}$ is the ith row of the array A.

1.1.4. *Procedures*

It is often expedient to give a name to a segment of a program which computes certain well-defined results and to refer elsewhere to this segment with functional notation. The names for these *procedures*—*functions* (a single result) or *routines* (several results)—may have the same form as variable names but, in addition, may be mnemonic titles formed from a few lower case letters. This name followed by an argument list is called a *heading* and is placed in the statement label column on line 0 of the segment. The arguments appear in parentheses, separated by commas, following the procedure name. A procedure thus has the following form:

A.0 *proc*$(a_1, a_2, \ldots)$
A.1   STATEMENT
A.2 #1 STATEMENT
.   .
.   .
.   .

A procedure referenced as a function may contain formulae such as

$$proc(a_1, a_2, \ldots) = E$$

which indicate that the value of the expression E is being assigned as the result (i.e. value) of the function. Such formulae appear as the last line of the segment or are followed by the statement

exit from procedure

(see § 1.3), which indicates that calculation of the functional value is complete.

Reference to a procedure is made in mnemonic functional notation, either within a formula in the case of a function, e.g.

$$y = fct(x,y,a)+2$$

or by means of a statement

execute *proc(a,b)*

in the case of reference to a routine. Reference to "generation procedures" is discussed in § 1.4.2.

Most often, the arguments appearing in a procedure reference represent values which are to be assigned at the time of execution as values of the corresponding argument variables whose names appear in the procedure heading. For sake of efficiency, however, it is sometimes better not to transmit the value of an array (i.e. the values of all of its entries), but to transmit only the *name* of the array (actually a pointer to information describing the array—see § 2.4.2), letting the procedure fetch or store entries in their original site. We place quotation marks about the name of a variable to indicate the name itself rather than the variable's value. This notation is used primarily for argument transmission, as illustrated in Fig. 1.2, but is sometimes useful in simple formulae such as

$$"A" = "B_{j,\,\_}" \quad \text{or} \quad "X" \leftrightarrow "Y".$$

These names do not constitute a distinct data-type—no operations other than the assignment operation (above examples) are defined with names as operands.

.
.
.

A.5   for $j \in J: M_j = \ldots$

A.6 $y = x + proc(a, "M")$

.   .
.   .
.   .

B.0   *proc(d, "A")*

.   .
.   .
.   .

$$z = \sqrt{(A_i)}$$

.
.
.

FIG. 1.2. Argument transmission of a name.

(Of course, we often use quotation marks within the text in the customary manner. However, even when used to delimit an out of context formula, there should be no confusion with either names or comments within a program.)

Argument transmission by name is frequently, but not exclusively, used merely to accommodate additional output from the procedure. To distinguish such usage we occasionally put strictly output arguments to the right of a semicolon in the argument list. Thus a heading

$$\text{P.0} \quad fct(x,\text{``}A\text{''};\text{``}Z\text{''})$$

has x as an input argument (transmitted by value), Z as a strictly output argument, and A as an input argument which might be altered within the procedure. Within the procedure itself, x will appear only on the right side of a formula, Z only on the left side, and A perhaps on both sides.

The data-type for arguments transmitted by name is carried along with the name. Ideally, the data-type for an argument transmitted by value can also be transmitted at "execute-time". However, practical considerations (see § 2.4.3) require that this data-type information be available at "compile-time" (so that procedures can be translated independently of references to them). Although in many cases the data-type of an argument can be discerned from context, there are times, due to admission of mixed expressions (see, for instance, § 1.2.1), when we must resort to declarations. These declarations appear on lines just following the heading, as in the following example:

D.0 $A^*(S,A,B,Z)$

D.1   (set) A, B; (real) x

D.2   (real array) Z

EXERCISES

1. Calculate the values of the following expressions given $j = 5, i = 8, \bar{r} = 7.1, T = -0.8$:

  (a) $(j! \pmod 3)+j^3 \pmod{i-1}$.
  (b) $|[-\bar{r}]|-[T]$.
  (c) $(2,3,4)\times(-1,7,6)-(j,3i,j-i)$.

2. Write a procedure for calculating $exp(x)$ using the approximation

$$exp(x) \approx \frac{12+6x+x^2}{12-6x+x^2}.$$

1.2. Set Manipulation

Finite sets of natural numbers, one of the basic data-types recognized in our language, play a fundamental role in combinatorial computing. The notation adopted for the manipulation of such data is essentially conventional mathematical notation (now taught in most elementary schools), although certain symbolism, and its associated data handling, does suggest computer orientation. These sets are basically handled as *unordered* sets, but we do occasionally take advantage of the natural ordering of their elements. (Note: To lessen confusion, we refer to "elements" of unordered sets and "entries" or "components" of arrays.)

1.2.1. *Notation for Sets*

As previously mentioned, the conventions adopted for naming variables is independent of the "values" (real numbers, sets of natural numbers, or arrays of these quantities) taken by the variable. Thus the statement

$$S = \{ \ \} \qquad \text{"}\{ \ \} \text{ is the void set"}$$

establishes S as the name of a set variable.

The curly bracket notation for explicit tabulation of a set is not used merely for constant sets. For example, $\{X_j\}$, $\{a-1,2a\}$, and $\{6,p,p^2\}$ are sets with variable elements. The natural number elements, which may be the computed values of general expressions, are listed within the braces, separated by commas. Note (as usual) the distinction between the natural number q and the *singleton set* $\{q\}$, the set whose single element is q.

Sets containing intervals of consecutive natural numbers, particularly sets of the form $\{0,1,2,\ldots,n-1\}$, occur frequently in our work, and it is convenient to have a notation to aid in their construction. Consequently, we interpret the natural number n, when it appears in set-theoretic context (i.e. within a parenthesis level of an expression which contains set-theoretic operations), as the set of natural numbers less than n, that is,

$$0 \equiv \{ \ \}$$
$$1 \equiv \{0\}$$
$$2 \equiv \{0,1\}$$
$$\cdot$$
$$\cdot$$
$$\cdot$$
$$n \equiv \{0,1,\ldots,n-1\}$$

In order to force set-theoretic context in expressions containing no operations, we use a parenthetical declaration. For example, the statement

$$A = (\text{set}) \ n$$

assigns the value $\{0,1,\ldots,n-1\}$ to the set variable A. In conjunction with basic set operations (next section), this interpretation greatly facilitates the formation of many sets.

The *cardinality* of a set—the number of elements in the set—is given by means of the "absolute value" notation, that is, $|S|$ represents the number of natural numbers in S. Note that this notation is consistent with our interpretation of a natural number as a set. (Although not used in this book, this notation could also be applied to ordered sets—1-arrays—to reveal an unknown coordinate range.)

1.2.2. *Operations*

The definitions of the primary set-theoretic operations are as follows:

$$
\begin{aligned}
a' &\equiv \{x : x \notin a\} && \text{complementation} \\
a \cup b &\equiv \{x : x \in a \quad \text{or} \quad x \in b\} && \text{union} \\
a \cap b &\equiv \{x : x \in a \quad \text{and} \quad x \in b\} && \text{intersection} \\
a \sim b &\equiv a \cap b' && \text{subtraction} \\
a \triangle b &\equiv (a \sim b) \cup (b \sim a) && \text{symmetric subtraction}
\end{aligned}
\tag{1.3}
$$

The notation for these operations is as shown; i.e., prime denotes complement, cup denotes union, etc. The operands are, of course, sets of natural numbers; this includes natural numbers interpreted as sets (e.g. $6 \sim 4 = \{4,5\}$). The "universe" is the set of all natural numbers ($\omega = \{0,1,2,\ldots\}$), hence the absolute complement of a finite set is an infinite set (see § 2.1); most frequently, however, relative complements—e.g. $n \sim a$, n a natural number—are used.

We note (in anticipation of later applications) that if the universe is restricted to $\{0\}$ ($= 1$), then these operations are isomorphic (under $1 \leftrightarrow$ "true" and $0 \leftrightarrow$ "false") to the Boolean operations of *not, or, and, and not,* and *exclusive or,* respectively. Also, *exclusive or (symmetric difference)* is addition modulo 2.

It is expedient to define notation for three additional binary operations upon sets. Let S be a set and let n be a natural number:

$$
\begin{aligned}
S \oplus \{n\} &\equiv \{x+n : x \in S\} \\
S \ominus \{n\} &\equiv \{x-n : x \in S \quad \text{and} \quad x-n \in \omega\} \\
\{n\} \ominus S &\equiv \{n-x : x \in S \quad \text{and} \quad n-x \in \omega\}
\end{aligned}
$$

For example, $\{0,6,11\} \oplus \{3\} = \{3,9,14\}$; $\{0,6,11\} \ominus \{3\} = \{3,8\}$; and $\{5\} \ominus \{1,3,5,8,9\} = \{0,2,4\}$. As we shall see in Chapter 2, in the implementation of sets as bit-patterns, these operations are effectively *translations* and *reflections* ("shift" operations) of those patterns. As the reader may discover, straightforward generalizations of these operations are possible.

We do not establish a hierarchy for the binary operations defined in this section. Explicit sequencing of the operations is indicated within an expression by use of the grouping symbols—parentheses and square brackets.

Notation for repetitive operations with sets is discussed in § 1.4.5.

1.2.3. *Functions Based on the Order of the Elements*

The ordering $0 < 1 < 2 < 3 < \ldots$ yields an ordering of the elements of any set of natural numbers. Thus it is meaningful to speak of the smallest, mth smallest, largest, or mth largest element of a set. Such reference is accomplished by means of a special

Set	Serial number	Binary representation of N
$s = set(N)$	$N = nbr(s)$	
{ }	0	...0000
{0}	1	...0001
{1}	2	...0010
{0,1}	3	...0011
{2}	4	...0100
{0,2}	5	...0101
.	.	.
.	.	.
.	.	.

FIG. 1.3. The $set(\ )$ and $nbr(\ )$ functions.

script notation. If S is a set, then $(S)_i$ is the ith smallest element of S and $(S)^i$ is the ith largest element of S. In both cases counting begins with one, so $(S)_1$ is the smallest and $(S)^1$ the largest element. For example, if $S = \{0,7,9,16,43,44\}$, then $(S)_1 = 0$, $(S)^3 = (S)_4 = 16$, and $(S)^1 = 44$. This notation is most often used when the desired element actually exists. However, the language does contain a special conditional statement (see §§ 1.3.2 and 2.1.4) enabling the programmer to determine if the element exists. Also it is convenient to define $(S)_0 = -1$.

The element ordering induces a natural ranking of all possible sets as shown in Fig. 1.3. With each "serial number" 0, 1, 2, ... we associate the set of exponents corresponding to existing terms in the binary expansion of the number and vice versa. We use the mnemonic notations "$nbr(s)$" and "$set(N)$" for computation of this association. For example, $nbr(\{0,2,3\}) = 13$ and $set(10) = \{1,3\}$. Actually this notation merely expresses the normal representation of sets of natural numbers (see § 2.1.1), hence the computer operations required to effect these functions are most often vacuous. Thus, when the use of the serial number of a set is clearly implied (e.g., in set comparison for ranking purposes—see § 1.3), the function mnemonic nbr will not be written. (This is equivalent to defining the arithmetic operations on our sets.)

EXERCISES

1. Tabulate the following sets:
   (a) $9 \sim 5$, $9 \oplus \{5\}$, $9 \sim \{5\}$, $\{9\} \ominus 5$.
   (b) $8 \cup 3$, $8 \cap 3$, $8 \cup \{8\}$, $8 \cap \{8\}$.
   (c) $\{7\}' \triangle \{6\}'$, $[(7 \oplus \{1\}) \triangle 6']'$.

2. For $S = \{2,3,8,11\}$, $(S)_{|S|} = ?$ $(S)^{|S|} = ?$

3. For a natural number x, $nbr(set(x)) = ?$ $nbr(x) = ?$

4. Let $S \subset \{0,1,\ldots,2^k - 1\}$ for an arbitrary positive integer k. Does $set(2^k - nbr(S) - 1) = 2^k \sim S$?

1.3. Transfer of Control—Conditional Statements

Apart from the extreme rapidity with which individual operations are performed, the power of modern electronic computing stems from the ability of the machine to discriminate on computed results and to alter the sequence of its calculation accordingly. Within the machine, such facility usually appears as various test and conditional and unconditional "transfer of control" (i.e. "branch" or "jump") instructions. In the language, this basic facility is accomplished, in the more elementary cases, by conditional and unconditional transfer of control statements.

1.3.1. *Basic Statement Structure*

An *unconditional transfer of control statement* is needed as a basic tool for altering the sequence of the calculation. Such facility is accomplished with the common (though disappearing) *go-statement*; for example, the statement

$$\text{go to} \; \# \; 6$$

indicates that the sequence of the calculation should continue with the statement which is labeled $\# 6$.

A *conditional statement* (or *test*) is a subordinate clause introduced by one of the conjunctions "if" or "unless" (or "while" or "until"—see § 1.3.4) and terminated by a colon. Between the conjunction and the colon stands a *condition*—a "mathematical sentence"—which at time of execution is either true or false. Following the colon there is one statement or more—the independent part or *consequent* of the *English* sentence —to be executed if the test is successful. The consequent appears either on the same line as the test or on successive lines indented slightly (to the right) from the horizontal position of the left edge of the test. In indented form a consequent may, of course, contain further indentation.

A simple example of the use of an *if-statement* (conditional statement with "if" as the conjunction) is as follows:

$$
\begin{array}{ll}
\text{T.1} & \text{if } A \geqslant 0: \\
\text{T.2} & \quad S = R \cup V \\
\text{T.3} & B = 4
\end{array}
$$

In this example, if at the time the test on line T.1 is executed the variable A has a non-negative value, then line T.2 will be executed next in sequence, followed by line T.3. On the other hand, if A has a negative value, then line T.2 will not be reached at this time, and calculation proceeds immediately with line T.3. Thus the consequent of an if-statement is entered only when the test is successful, i.e. when the condition is true. The above example could also have been written

$$
\begin{array}{l}
\text{if } A \geqslant 0: \; S = R \cup V \\
B = 4
\end{array}
$$

The use of the conjunction "unless" serves to negate the condition. For example, the statement

$$\text{unless } A = 0: \; x+1 \to x$$

is equivalent to

$$\text{if } A \neq 0: \; x+1 \to x,$$

where x will be increased by one *unless* A is equal to zero. The consequent of an unless-statement is entered only when the condition is false. This terminology is of particular use when the condition is quite involved (see later discussion) and difficult to negate "by hand". (An alternative used in other languages is the use of the word "not" or other Boolean complementary symbol to indicate negation of a parenthesis level.)

1.3.2. *Simple Conditions*

The examples of the previous section illustrate the use in simple conditions of the symbols $\geqslant$, $=$, and $\neq$ for "greater than or equal to", "equal to", and "not equal to" respectively. In general, the symbols

$$=, \; \neq, \; <, \; \not<, \; >, \; \not>, \; \leqslant, \; \not\leqslant, \; \geqslant, \; \not\geqslant$$

are employed to express the common binary comparisons of real numbers (and their negations). A "modular" comparison is denoted with the identity symbol and a "(mod n)" phrase. For example,

$$\text{if } a \equiv b \; (\text{mod } p):$$

asks if the whole numbers a and b are congruent modulo p (i.e. if $a-b$ is divisible by p). The important special case of parity testing has its own notation, namely

$$\text{if } a \text{ is even:} \qquad \text{and} \qquad \text{if } a \text{ is odd:}$$

The basic set-theoretic comparisons include tests for equality and for inclusion (containment) and their negations expressed by the following symbols:

$$=, \; \neq, \; \subset, \; \not\subset, \; \supset, \; \not\supset.$$

For example, the test

$$\text{if } S \subset T:$$

asks whether every element of S is also an element of T. (We consider the condition "$S \subset S$" true, hence have no need for the comparators $\subseteq$ and $\supseteq$. The seldom-used test for *proper* inclusion, i.e. $S \subset T$ and $S \neq T$, would be handled by a compound condition—see § 1.3.3.) Element inclusion conditions use the symbols $\in$ and $\notin$, which are read "is an element of" and "is not an element of", respectively; for example,

$$\text{if } i \in I:$$

asks whether the natural number i is an element of the set I. The test

$$\text{if nonexistent:}$$

may be used immediately following a statement containing a single occurrence of the notation $(S)_j$ or $(S)^j$; it is successful when the requested element does not exist. (This test may also be used with the "first..." notation described in § 1.4.5.)

When applied with sets as comparands, the arithmetic comparators imply comparison of the serial numbers of the sets. For example, with

$$S = \{0,4,5\} \quad \text{and} \quad T = \{1,6\}$$

the condition

$$S < T$$

is true since $nbr(S) = 49$ is less than $nbr(T) = 66$. The arithmetic comparators may also be used with arrays as comparands. This condition and others involving search through an array of reals or sets are discussed in § 1.4.5.

All conditions used in this book consist of comparisons between real, set, or array quantities, which, by the way, may be given by arbitrary mathematical expressions. We do not, as has been done in some languages (notably Algol), introduce a Boolean data-type, allowing variables which assume only the values "true" or "false" to substitute for conditions. The familiar case where the primary purpose of a procedure calculation is the selection of one of two alternatives is handled by using the numbers 0 or 1 as possible function results. The test then might take the form

$$\text{if } fct(\text{ARGUMENTS}) = 1:$$

for example. (Variables which assume only two values, hence which may be represented by a single bit, are sometimes called "toggles".)

The simple go-statements and if-statements presented so far, although logically sufficient for controlling program flow, do not possess the power or sophistication which we feel the scientist uninterested in programming detail deserves. However, these statements can and will be used as building blocks in the definitions of more elaborate language for program control.

1.3.3. *Compound Conditions*

Simple conditions may be joined with the conjunctions "and" and "or" to form compound conditions. When "and" is used, the consequent of the test is entered (in the case of an if-statement) only when *all* the simple conditions are true. For example,

$$\text{if } A = 0 \quad \text{and} \quad x \leqslant 3:$$
$$\text{CONSEQUENT}$$

is equivalent to

$$\text{if } A = 0:$$
$$\text{if } x \leqslant 3:$$
$$\text{CONSEQUENT}$$

When "or" is used, the consequent of an if-statement is entered when *any* one of the simple conditions is true. For example,

$$\text{if } i \in I \quad \text{or} \quad a \subset b:$$
$$\text{CONSEQUENT}$$
$$\text{PROGRAM}$$

is equivalent to

$$\text{if } i \in I:$$
$$\# 1 \qquad \text{CONSEQUENT}$$
$$\text{go to } \# 2$$
$$\text{if } a \subset b: \quad \text{go to } \# 1$$
$$\# 2 \quad \text{PROGRAM}$$

Considering the "and" operation as logical multiply and the "or" operation as logical add, using parentheses for logical grouping (as well as for arithmetic grouping), and commas to substitute occasionally for the words "and" and "or", quite involved compound conditions may be tested in a conditional statement. Although logical multiply can be acknowledged to take precedence over logical add (as is customary in analogous arithmetic notation), we generally indicate specific grouping by extensive use of parentheses. An example of a compound conditional statement is

$$\text{unless} \quad (i \notin A, \; x = 0 \quad \text{and} \quad a \subset b) \quad \text{or} \quad A \cap B \neq 0: \qquad (1.4)$$

The following shorthand notations are occasionally used to express the two compound conditions of "implication" and "equivalence". The test

$$\text{if } C_1 \Rightarrow C_2:$$

means

$$\text{if } C_2 \quad \text{or} \quad (\bar{C}_1 \text{ and } \bar{C}_2):$$

where C_1 and C_2 are arbitrary conditions and a bar indicates negation of the condition, and the test

$$\text{if } C_1 \Leftrightarrow C_2:$$

means

$$\text{if } (C_1 \text{ and } C_2) \quad \text{or} \quad (\bar{C}_1 \text{ and } \bar{C}_2):$$

1.3.4. *Additional Forms*

The frequency of occurrence of the program construction shown in segment *A* of Fig. 1.4 makes a more suggestive notation desirable. This form occurs when PROGRAM 1 alone is to be executed when the CONDITION is true and PROGRAM 2 alone is to be executed when the CONDITION is false. It may be written, more symmetrically, as shown in segment B of Fig. 1.4, or in a one-line form with a semicolon separating

```
       A.1        if CONDITION:
        ⋮            PROGRAM 1
                    go to # 1
                  PROGRAM 2
       # 1        PROGRAM 3
       B.1        if CONDITION:
        ⋮            PROGRAM 1
                  otherwise:
                  PROGRAM 2
                  PROGRAM 3
```

FIG. 1.4. Meaning of "otherwise" statement.

PROGRAM 1 from the word "otherwise", i.e.

if CONDITION: PROGRAM 1; otherwise: PROGRAM 2

Two other frequently appearing forms suggest the use of different introductory conjunctions. We use the conjunction "while" in place of "if" when the condition is to be re-tested after calculation of the consequent, with re-calculation of the consequent *as long as* the condition is true. Thus the form

```
          while CONDITION:
               CONSEQUENT
          PROGRAM
```

is equivalent to

```
      # 1   if CONDITION:
                 CONSEQUENT
               go to # 1
          PROGRAM
```

Similarly, we use the word "until" in place of "unless".

EXERCISES

1. Rewrite the program given by (1.4) using tests containing simple conditions only.

2. (a) Write a test for the disjointness of two sets A and B.

   (b) What relationship do sets A and B have (e.g. draw a Venn diagram) if the condition "$A \cap B \neq 0$ and ($A \not\subset B$ or $B \not\subset A$)" is true?

3. Under what conditions are the tests "if $a = b \pmod{p}$:" and "if $a \equiv b \pmod{p}$:" not equivalent?

4. For which of the following tests would PROGRAM 1 be executed?

   A.1 if $n \subset n-1$: PROGRAM 1   "n is a natural number."

   B.1 unless $3 \in S \leftrightarrow \{3\} \subset S$: PROGRAM 1   "$S$ is a set."

   C.1 while $T < \{3,6\}$:   "$T = \{5,6\}$ initially."

   $T \ominus \{1\} \rightarrow T$; PROGRAM 1

   D.1 if $(S)^1 \in S \oplus \{1\}$: PROGRAM 2; otherwise: PROGRAM 1.

5. Using only the notation introduced so far, write a program to locate the maximum entry in the 1-array of reals $R_{0 \text{ to } n-1}$.

6. Write a procedure for calculating the greatest common divisor of the integers a and b. The greatest common divisor of a and b, $gcd(a,b)$, can be calculated from *Euclid's algorithm* [34], i.e. for $a \geqslant b > 0$ it is the last nonzero remainder r_n in the sequence:

$$a = q_1 b + r_2, \qquad q_1 = \left[\frac{a}{b}\right] \qquad \text{hence} \qquad 0 \leqslant r_2 < b$$

$$b = q_2 r_2 + r_3, \qquad q_2 = \left[\frac{b}{r_2}\right] \qquad \text{hence} \qquad 0 \leqslant r_3 < r_2$$

$$r_2 = q_3 r_3 + r_4, \qquad q_3 = \left[\frac{r_2}{r_3}\right] \qquad \text{hence} \qquad 0 \leqslant r_4 < r_3$$

$$\vdots \qquad\qquad\qquad\qquad\qquad\qquad\qquad \vdots$$

$$r_{n-2} = q_{n-1} r_{n-1} + r_n \qquad\qquad\qquad 0 \leqslant r_n < r_{n-1}$$

$$r_{n-1} = q_n r_n.$$

1.4. Notation for Iteration and Recursion

Most computer algorithms contain in their specification instructions to repeat a certain job, each time with different values of the variables involved. A simple example is the calculation of the *Fibonacci sequence*, $u_0, u_1, \ldots, u_N$, from the recurrence relation

$$u_n = u_{n-1} + u_{n-2}, \quad u_0 = u_1 = 1. \tag{1.5}$$

A computer algorithm for performing this calculation might say "Let u_0 and $u_1 = 1$; then, for $n = 2, 3, \ldots, N$, calculate $u_n = u_{n-1} + u_{n-2}$". It is desirable to have in our language a compact and natural notation for expressing these iterative calculations. Although our language makes no such distinction, it is advantageous for textual descriptions to distinguish between "recursion" in which successive calculations depend on previous calculations so that the order of indexing is important (Fibonacci example) and "iteration" in which each calculation of the sequence is independent of every other calculation so that the order of indexing is unimportant (setting $A_i = 2i$, for $i = 1$ to I, for example). The more involved case of "recursive procedures", parameter dependent functions whose evaluation depends on the evaluation of the same function with different parameter value, is discussed in § 1.5.

1.4.1. *For-statement Structure*

An *iteration* or *recursion* or *loop* is indicated in our language by an *iteration quantifier*—a *for-clause*—giving the range of the iteration, and the *body of the loop*, the program to be repeated. The iteration quantifier is a subordinate clause introduced by the conjunction "for" and terminated by a colon. The *index range*, which stands between the word "for" and the colon, indicates by means of an arithmetic or set-theoretic sequence,

PROGRAM

I.1	$u_0, u_1 = 1$
I.2	for $n = 2, 3, \ldots, N$:
I.3	$u_n = u_{n-1} + u_{n-2}$

PROGRAM

FIG. 1.5. Recursion notation sample.

PROGRAM

R.1		$u_0, u_1 = 1$
R.2		$n = 2$
R.3		go to $\#2$
R.4	$\#1$	$n+1 \to n$
R.5	$\#2$	if $n > N$: go to $\#3$
R.6		$u_n = u_{n-1} + u_{n-2}$
R.7		go to $\#1$
R.8	$\#3$	PROGRAM

FIG. 1.6. Recursion notation resolution.

a simple recurrence relationship, an index set, a procedure calculation, or by implication, successive values to be assigned to a dummy (loop) index. As with consequents of conditional statements, the body appears either on the same line or indented on following lines. Also, the body of the loop, being an arbitrary program, may contain further indentation. As a simple example of recursion specification, the Fibonacci sequence calculation of (1.5) could be written as shown in Fig. 1.5. An equivalent program written without advantage of the loop notation is shown in Fig. 1.6. This illustrates a *resolution*; in this case a resolution of the iteration notation when the index range is given as an arithmetic sequence. Note that if N is less than 2, the body of the loop (line I.3 of Fig. 1.5) would not be executed—see lines R.3 and R.5 of Fig. 1.6. Note, also, that the index n has the value $N+1$ when the recursion is complete (provided $N > 0$, of course). Since the body of the loop is such a simple program, a one-line form, e.g.

$$u_0, u_1 = 1; \text{ for } n = 2, 3, \ldots, N: u_n = u_{n-1} + u_{n-2}$$

could be used in this case.

1.4.2. *Simple Range Specification and Generation Procedures*

The index range for a loop is specified in one of several ways. Most common is to specify an arithmetic sequence of values which the dummy index is to successively adopt, by giving the first, second, and last terms of the sequence, as in the following examples:

$$\begin{aligned}
&\text{for } i = 0, 1, \ldots, 20: \\
&\text{for } A = I, I-1, \ldots, 0: \\
&\text{for } j_i = k+1, k+3, \ldots, K-2: \\
&\text{for } x = 3y, 3y+d, \ldots:
\end{aligned} \qquad (1.6)$$

The last example shows that the last term of the sequence need not be given (exit from the loop is handled by other means). Note that the order in which values are to be assigned to the loop index (especially important in recursive work) is implied by the range specification. If the increment (i.e. the constant difference of the arithmetic sequence) is $+1$, then the form

$$\text{for } i = A \text{ to } B:$$

is sometimes employed.

A more general way to specify the successive values of the index is to give the first value and then a first-order recurrence formula for calculating a new value from the previous value; for example, the quantifier

$$\text{for } i = 1, \ldots, 2i \to i, \ldots: \qquad (1.7)$$

assigns the values 1, 2, 4, 8, 16, ... to the index i. In this notation, the extent of the iteration is controlled from within the loop or by attaching an until-clause between the final three dots and the colon (see § 1.4.4). Also, it is sometimes convenient to omit the index initialization from the for-statement as in

$$\text{for } \ldots, 3j+2 \to j, \ldots:$$

for example; this case occurs most frequently in compound range specifications (also see § 1.4.4).

We sometimes use a "set-theoretic sequence" notation which is quite similar to the arithmetic sequence notation. Instead of indicating the second term (hence successive terms) by an addition or subtraction of an increment, the second term is formed by applying a set-theoretic operation to the first term and the "increment", for example:

$$\begin{aligned}
&\text{for } A = B, B \cup C_i, \ldots: \\
&\text{for } S = s, s \oplus \{3\}, \ldots:
\end{aligned} \qquad (1.8)$$

The extent of the iteration is usually controlled by means of an until-clause or "coincident" range specification. Note that, to be reasonable, a sequence formed from the union operation requires a succession of increments.

Another form for specifying an index range is by use of an index set, as in the examples

$$\begin{aligned}
&\text{for } j \in J: \\
&\text{for increasing } j \in J: \\
&\text{for decreasing } j \in J:
\end{aligned} \qquad (1.9)$$

which state that the dummy index j assumes for its values each element of the set J. The natural number values are assigned in increasing order unless the modifier "decreasing" appears before the index name as in the third example. The form

$$\text{for } x \subset y: \tag{1.10}$$

is also used; it states that the dummy set x assumes for its values each subset of the set y in increasing order of serial number.

We allow components of a range specification (e.g. an upper limit or an index set) to vary within an iteration. In some cases this shifts the burden of efficient translation of the source program into machine code onto the programmer (see § 2.3.1), but the increased flexibility compensates for this.

A still more general form for index range specification allows a "generation procedure" to assign values to the dummy index. An example is

$$\text{for each } \textit{factor}(n) \to x: \text{BODY} \tag{1.11}$$

where the procedure $\textit{factor}(\ )$ successively produces results which are assigned as values of the dummy index x. This form will sometimes be used with no dummy index specified:

$$\text{for each } \textit{fct}(a,b): \text{BODY}$$

The implementation of iteration quantifiers involving index sets or generation procedures is slightly different from that involving arithmetic sequences. The resolution of the loop

$$\text{for } i \in I: \text{BODY} \tag{1.12}$$

into a form involving only formulae and conditional and unconditional transfer of control statements is exhibited in Fig. 1.7. In this program, $\textit{next}(\ )$ is the name of a (library) function which produces the next element of the set I given that set and the previous element i. This procedure must also distinguish between the first entry and successive entries, and must indicate to the main program when the iteration is complete. This is done by use of the toggles $\textit{Entry}$ and $\textit{Exit}$, shown in Fig. 1.7 as normal variables.

		PROGRAM
I.1		$\textit{Entry} = 1$
I.2		go to #2
I.3	#1	$\textit{Entry} = 0$
I.4	#2	$\textit{next}(i, I, \textit{Entry}, \text{"Exit"}) \to i$
I.5		if $\textit{Exit} = 1$: go to #3
⋮		BODY
		go to #1
	#3	PROGRAM

Fig. 1.7. Resolution of index set controlled iteration.

This testing and setting of special toggles distinguishes this function as a *generation procedure*.

The iteration of (1.11), which uses explicit reference to a generation procedure, has a similar resolution; the line

$$\text{I.4} \quad \#2 \quad factor\,(n,\text{``}x\text{''}) \rightarrow x$$

replaces line I.4 of Fig. 1.7. The references to *Entry* and *"Exit"* as arguments are omitted here, and, due to their special nature (see § 2.1.4), are also omitted from procedure headings. However, within generation procedures themselves (see Chapter 5 in particular), we express the testing and setting of *Entry* and *Exit* in normal notation.

The last basic form for index range specification which we use assigns values to the loop index "by implication". In this case a range is not explicitly given in the quantifier; only the name of the index is given. Presumably this index appears within the body of the loop as a subscript upon an array name. The current coordinate range of that dimension of the array is used for the index range of the iteration. For example, when A is a known 1-array with subscript range 0 to n, the iteration

$$\text{for } i\colon A_i = i^2 \tag{1.13}$$

is equivalent to

$$\text{for } i = 0 \text{ to } n\colon A_i = i^2$$

This notation is used only when there is no ambiguity as to what the range should be. Such a quantifier may be implemented in a fashion similar to one involving an index set or generation procedure. For example, the resolution of (1.13) is given by Fig. 1.7 with the I of line I.4 replaced by A; it is assumed that transmission of an array as an argument includes transmission of its subscript ranges.

1.4.3. *Auxiliary Transfer of Control Statements*

It is sometimes necessary (although it often reflects either a defect or misuse of the language) to be able to transfer control from within the body of a loop to either the incrementing part of the loop control (statement $\#\,1$ of Figs. 1.6 and 1.7) or to the point of exit from the loop (statement $\#\,3$ of the figures). This is accomplished by use of the statements

reiterate

and

exit from loop

respectively. Such statements apply, of course, to the loop "immediately containing" them.

1.4.4. *Compound Range Specification*

There are several ways in which index ranges may be compounded. They are best described by showing (in Fig. 1.8) the resolution, hence implementation, of each of the following sample notations:

S.   for $i \in I$ such that i is even: BODY

T.   for $i = 1, 2, \ldots$ until $k_i \equiv 0 \pmod{p}$: BODY

U.   for $i \in I$, as $j = 0, 1, \ldots, J-1$: BODY

V.   for $i = 1$ to 10, and then $i = 12, 14, \ldots, 2j$: BODY

W.   for $i \in I; j \in J$: BODY.

Segment S illustrates a compact notation for further restricting the values of the loop index for which the iteration is performed. We occasionally use the backwards epsilon symbol $\ni$ in place of the words "such that" to condense the notation still more. Segment T illustrates the manner in which additional, or essential, conditions for termination of the loop may be stated.

S.1   for $i \in I$:

S.2   unless i is even: reiterate

⋮   BODY

T.1   for $i = 1, 2, \ldots$:

T.2   if $k_i \equiv 0 \pmod{p}$: exit from loop

⋮   BODY

U.1 $Entry = 0; j = 0$

U.2   go to #2

U.3   #1 $Entry = 1; j+1 \rightarrow j$

U.4   #2 $next(i,I) \rightarrow i$

U.5   if $Exit = 1$ or $j > J-1$: go to #3

⋮   BODY
          go to #1

     #3   NEXT STATEMENT

V.1   for $i = 1$ to 10: BODY

V.2   for $i = 12, 14, \ldots, 2j$: BODY

W.1   for $i \in I$:

W.2   for $j \in J$:

⋮   BODY

FIG. 1.8. Resolutions of compound iterations.

The form of segment U, illustrating *coincident* index range specification (i.e. the two indices run simultaneously as the iteration proceeds), is extremely useful. Note the order in which the index initialization (line U.1 of Fig. 1.8) incrementing (line U.3) and testing (line U.5) is performed; it is often expedient to have the second index depend on the first—see (1.14) for an example. A range specification may contain more than two coincident indices. In all cases, exit from the loop occurs when any index has exhausted its range.

Segment V illustrates one method of denoting a changing increment or, in general, a succession of ranges for a single index; we call this *sequential specification*. The second "$i =$" is usually omitted if the second index is the same as the first. Segment W exemplifies a compact *product range* notation. Note that such multiple iterations, although "nested", require only a single indentation. Of course, a product range cannot be used when computation appears within the body of the "outside" loop but not within the "inside" loop.

These notations may be combined as in a statement of the form

for RANGE1 $\ni$ CONDITION1, as RANGE2 until CONDITION2; RANGE3:

for example. Resolution of such a quantifier proceeds in inverse order of the forms as presented in Fig. 1.8: one first resolves with respect to semicolons, then with respect to "and then" conjunctions, etc. Complex constructions involving parenthesis levels of basic forms are not required in this book.

1.4.5. *Notations Involving Iteration*

The various forms of index ranges are used with notations other than for-statements. Notations for summation, product formation, repeated union, intersection and symmetric subtraction, and for locating maxima and minima, are fashioned from the symbols

$$\Sigma, \Pi, \cup, \cap, \Delta, MAX, \text{ and } MIN$$

respectively by placing a range specification as a subscript on the respective symbol. For example, the sum of the first $N+1$ terms of the series expansion of e^x, namely,

$$1 + x + \frac{x^2}{2!} + \frac{x^3}{3!} + \cdots + \frac{x^N}{N!}$$

could be written, naturally, as

$$\Sigma_{n=0 \text{ to } N} \left(\frac{x^n}{n!} \right)$$

although, more efficiently, as

$$\Sigma_{n=0 \text{ to } N, \text{ as } t=1, \ldots, tx/n \to t, \ldots} (t) \tag{1.14}$$

The resolution of these notations is straightforward. In terms of previously defined notation, the expression

$$S = \cup_{i \in I} (A_i \cap B_i) \sim T$$

$$Union = \{\ \}$$
$$\text{for } i \in I:$$
$$Union \cup (A_i \cap B_i) \rightarrow Union$$
$$S = Union \sim T$$

FIG. 1.9. Resolution of union notation.

would appear as in Fig. 1.9 for instance, where *Union* is the name of a dummy variable introduced by the translator. With singleton sets as operands, this union notation is frequently employed for set formation; for instance,

$$\bigcup_{j \in J \ni j \equiv 0 \ (\text{mod } n)} (\{j\})$$

is the subset of *J* whose elements are divisible by *n*.

Range specifications are also used as subscripts on the symbols

$$\exists \quad \text{and} \quad \forall$$

to create the *existential* and *universal quantifiers*. These quantifiers are used in tests of the form

$$\text{if } \exists_{\text{RANGE}} \ (\text{CONDITION}): \text{CONSEQUENT}$$

and

$$\text{if } \forall_{\text{RANGE}} \ (\text{CONDITION}): \text{CONSEQUENT}$$

which resolve as shown in Fig. 1.10. The condition within the parentheses may itself be a

"existential resolution"

E.1	$Dummy = 0$
E.2	for RANGE:
E.3	if CONDITION:
E.4	$Dummy = 1$; exit from loop
E.5	if $Dummy = 1$:
E.6	BODY

"universal resolution"

U.1	$Dummy = 1$
U.2	for RANGE:
U.3	if CONDITION: reiterate
U.4	$Dummy = 0$; exit from loop
U.5	if $Dummy = 1$:
U.6	BODY

FIG. 1.10. Resolutions for existential and universal quantifiers.

quantified condition or part of a compound condition. Also, a quantified condition, treated as a simple condition, may be part of a compound condition. Thus extremely complex conditions may be written with little pencil; their resolution to "simpler", but more long-winded, forms may always be derived inductively from the basic resolutions illustrated in Fig. 1.10. An example of such a test is

$$\text{if } a \subset S \text{ or } \forall_{x \in |X} (x \neq z \text{ and } \exists_{p \in P} (p \in A_x)): \qquad (1.15)$$

However, since the human mind generally balks at about three levels, extremely complex conditions usually appear in our algorithms partly resolved.

It is convenient to have a notation for locating the "first" value of an index satisfying a given condition. The notation we use is closely related to the existentially quantified condition notation. For example, the formula

$$j = (\text{first } k \in K \ni A_k = k) + 2$$

is equivalent to

$$\text{if } \exists_{k \in K} (A_k = k): j = k + 2$$

(where, as before, the symbol $\ni$ substitutes for the words "such that"). Of course, in general, any range specification, simple or compound, may appear as the quantifier following the word "first", and indeed the order specified by that range is often significant. We use this notation when we are quite sure that such a value for the index does indeed exist. If not assured of existence, we may either employ the "if nonexistent" test described in § 1.3.2 or use an alternate notation, one involving the existential quantifier and the word "otherwise".

It is convenient to view two arrays as equal if their corresponding entries are equal. Thus if A and B are two arrays with subscript range 0 to $n-1$, say, the test

$$\text{if } A = B:$$

is equivalent to the test

$$\text{if } \forall_{i=0 \text{ to } n-1} (A_i = B_i):$$

Similarly, we find extensive use for the test

$$\text{if } A < B:$$

defined to be equivalent to

$$\text{if } \exists_{i=0 \text{ to } n-1} (A_i \neq B_i) \text{ and } A_i < B_i: \qquad (1.16)$$

which asks whether the first entry of A which is not equal to the corresponding entry of B is less than that entry of B. Note that the test is unsuccessful if the arrays are equal. Actually, for this test the subscript ranges of the arrays A and B need not be identical, the range of the existential quantifier is the same as that for A but not necessarily for B. The entries of the arrays may be sets as well as real numbers (see § 1.3.2).

EXERCISES

1. Write programs involving only formulae and transfer of control statements equivalent to the following statements:

(a) for $i = n, \ldots, [i/2] \to i, \ldots$ until $i = 1$: BODY.

(b) for $i = 0$ to $I-1 \ni A_i \neq \{ \ \}$ until $|A_i| > K$: BODY.

2. Give reasonable definitions for the following notations (i.e. resolve in terms of notation explicitly presented in the text):

(a) for RANGE except when CONDITION: BODY

(b) for RANGE while CONDITION: BODY

(c) unless $\exists_{\text{RANGE}}$ (CONDITION): CONSEQUENT

(d) for any $i = 1, 2, \ldots \ni$ CONDITION: BODY

(e) $S =$ all $i \in I \ni$ CONDITION

(f) $\#_{i \in I} (A_i = B_i)$.

3. For the arrays $C_{1 \text{ to } n, 1 \text{ to } m}$ and $D_{1 \text{ to } n, 1 \text{ to } m}$ resolve the test "if $C > D$:" in terms of compound conditional statements involving the existential quantifier.

4. Resolve the following statements in terms of formulae, for-statements and transfer of control statements, introducing any dummy variables needed:

(a) $a = MAX_i(MIN_j(A_{i,j}))$

(b) $N = \left| \bigcap_{z \in Z, \text{ as } q \in Q} (B_z \triangle C_q) \right|$

(c) The statement (1.15).

5. Assuming $A = (3,6,9,12)$, what is the value of j upon completion of each of the following lines of program?

(a) if $\exists_{i=0 \text{ to } 3}(A_i$ is even): $j = i$; otherwise: $j = -1$

(b) if $\forall_i (A_i \equiv 0 \pmod 3)$: $j = i$; otherwise: $j = -1$

(c) $k = \left[\text{first } j = 3,2,\ldots,0 \ni \exists_{i=0, 1, \ldots} (A_j = i^2) \right] - 1$.

6. Write a program for calculating the binomial coefficients $C_{n,r}, \ 0 \leqslant n \leqslant N, 0 \leqslant r \leqslant [n/2]$ from the recurrence relation $C_{n,r} = C_{n-1,r} + C_{n-1,r-1}$; $C_{n,0} = 1$. (Recall [36] that $C_{m,r} = C_{m, m-r}$.)

7. What function of the set variable S does $\sum_{i \in S} (2^i)$ produce?

8. Under what conditions (on A_- and B_-) are the tests

$$\text{if } \exists_{i \in I} (A_i = 0) \text{ and } \exists_{j \in J_i} (B_j = 0)$$

and

$$\text{if } \exists_{i \in I} \big((A_i = 0) \text{ and } \exists_{j \in J_i} (B_j = 0)\big)$$

equivalent?

9. As can be seen from Fig. 1.9, the value of $\bigcup_{i \in I} (S_i)$ is $\{ \ \}$ when I is the void set. Give natural definitions for $\sum_{i \in I} (N_i)$, $\Pi_{i \in I} (N_i)$, $MIN_{i \in I} (N_i)$, $MAX_{i \in I} (N_i)$ when $I = \{ \ \}$.

*10. If $a > b$, a is odd, and b is odd, then $gcd \ (a,b) = gcd(b,(a-b)/2)$. Incorporate this fact, as well as the fact that $gcd(a,b) = gcd(a,a-b)$ for all a and b, in an efficient program for the Euclidean algorithm [52] (see exercise 6 of § 1.3), using the iteration notation of this section.

1.5. Nested Iteration and Recursive Programming

Since the body of a loop may itself contain additional loops, multiple (but fixed) dimensional iteration may be expressed in our language by use of several levels of indentation. In combinatorial computing it often happens that iterations must be "nested" to a variable or indefinite depth. In fact, a basic tool of combinatorial computing —"backtracking"—requires just such a facility. Moreover, there is a close relationship between variably nested recursions and "recursive operations", operations defined in terms of themselves. In fact, such "recursive programming" is an increasingly important concept of advanced computing. Notation for variably nested iterations, hence backtracking and recursive programming, is based on a special control statement, a "with-statement".

The reader is advised not to expect to understand the intricacies of with-statements and their use from reading this section. The relevant language is presented here with little motivation. More discussion and many examples of nests and recursive programming are given in subsequent chapters, particularly Chapter 4.

1.5.1. *With-statement Structure*

A *nest* is a sequence of iterations, each one contained in (nested within) the body of the previous one, written as a single, parameter dependent iteration—the *nest iteration*—under control of the *nest quantifier* (or with-clause) indicating the extent of the nesting. The nest quantifier has the same form as an iteration quantifier with the exception that the word "with" is used in place of the word "for". The indented nest iteration, bracketed by special slanting three-dot symbols, follows the with-clause. As a simple example of a nest, consider the task of forming all k component vectors $(d_0, d_1, \ldots, d_{k-1})$ with $0 \le d_i < b$, $i = 0$ to $k-1$ (i.e. counting with base b). Segment A of Fig. 1.11 shows a program for this calculation when $k = 3$; segment B illustrates the nest form of this calculation, for *arbitrary k*.

A generic form for a nest is given in Fig. 1.12. The iteration and programs bracketed by each pair of three-dot symbols are to be considered repeatedly written, with appropriate substitution of the with-clause dummy index, in position indicated by the direction of the

"$k = 3$"

A.1   for $d_0 = 0$ to $b-1$:

A.2   for $d_1 = 0$ to $b-1$:

A.3   for $d_2 = 0$ to $b-1$:

A.4   USE (d_0, d_1, d_2)

"arbitrary k"

B.1   with $i = 0$ to $k-1$:

B.2 $\cdot\cdot\cdot$

B.2   for $d_i = 0$ to $b-1$:

B.3 $\cdot\cdot\cdot$

B.4   USE $(d_0, d_1, \ldots, d_{k-1})$

FIG. 1.11. Sample nests.

N.1   NEST QUANTIFIER "e.g. with $i = 1, 2, \ldots, I$:"

N.2 $\ddots$

N.3   PROGRAM 1

N.4   ITERATION QUANTIFIER "e.g. for $j_i = 0, 2, \ldots, J_i$:"

N.5   PROGRAM 2

N.6 $\ddots$

N.7   PROGRAM 3

N.8 $\cdot^{\cdot}$

N.9   PROGRAM 4

N.10   PROGRAM 5

N.11 $\cdot^{\cdot}$

N.12   PROGRAM 6

FIG. 1.12. The general nest notation.

dots, once for each value of that index. Thus each iteration is preceded by a PROGRAM 1; it contains within its body a PROGRAM 2, the more deeply nested iterations (lines N.6 through N.8), and a PROGRAM 4; it is then followed by a PROGRAM 5. During the execution of a nest, PROGRAM 2 is entered just before deeper nesting and PROGRAM 4 is entered when the deeper nesting is complete; PROGRAM 1 is executed first as each level is reached (note that PROGRAM 2 for an iteration immediately precedes the PROGRAM 1 for the next iteration). When an iteration is complete and return is made to the indentation level of the iteration quantifier, PROGRAM 5 is then entered just prior to "reiteration" in the containing loop. The *center of the nest*, PROGRAM 3, is entered when no further nesting is called for by the next quantifier. For notational brevity, we usually omit the three-dot symbols on lines N.8 and N.11.

A resolution of this notation (using the sample quantifiers given as comments in Fig. 1.12) in terms of formulae and transfer of control statements is shown in Fig. 1.13. Note (see lines R.2 and R.4) that if I is initially less than 1, there are no iterations to perform and control passes immediately to PROGRAM 3 and soon afterwards to PROGRAM 6. Note, also, that i has the value $I+1$ when PROGRAM 3 is entered (provided $I \geqslant 0$) and the value 0 when the calculation described by the nest is complete.

Most recursive operations do not require the generality allowed by Fig. 1.12 where program may appear at quite arbitrary points in the nest. In fact, since PROGRAM 2 of one iteration abuts PROGRAM 1 of the next iteration, as do PROGRAMS 4 and 5, there is some room for personal preference in writing variably nested iterations.

1.5.2. *Range Specification*

Not all of the simple index range constructions appropriate for iteration control can be used for nest control since a nest index moves forwards *and* backwards—the operation performed to generate successive index values must be invertible. Such is

R.1		$i = 1$
R.2		go to #2
R.3	#1	$i+1 \to i$
R.4	#2	if $i > I$: go to #7
R.5		go to #4
R.6	#3	$i-1 \to i$
R.7		if $i < 1$: go to #10
R.8		go to #8
R.9	#4	PROGRAM 1
R.10		$j_i = 0$
R.11		go to #6
R.12	#5	$j_i+2 \to j_i$
R.13	#6	if $j_i > J_i$: go to #9
R.14		PROGRAM 2
R.15		go to #1
R.16	#7	PROGRAM 3
R.17		go to #3
R.18	#8	PROGRAM 4
R.19		go to #5
R.20	#9	PROGRAM 5
R.21		go to #3
R.22	#10	PROGRAM 6

Fig. 1.13. Resolution of nest notation.

certainly the case for arithmetic sequence specification, the most common form of nest control. Certain set-theoretic operations are invertible (e.g. the union of disjoint sets) and are used to generate sequences of index values for nest control. The recurrence form (1.7), which would require inversion of an arbitrary expression, is not used.

The index set form (1.9) and generation procedure form (1.11) are occasionally employed. When so used the generation procedures themselves must be able to decrease (as well as increase) the nest index incrementally. This requires an extra entry toggle to discriminate backward from forward movement in the nest. The resolution of the statement

$$\text{with } proc(\ ) \to i$$

R.1 $Entry1 = 1; Entry2 = 0$
R.2   go to #2
R.3 $Entry1 = 0; Entry2 = 0$
R.4 $proc(\ ) \to i$; if $Exit = 1$: go to #7
R.5   go to #4
R.6   #3 $proc(\ ) \to i$
R.7   if $Exit = 1$: go to #10
R.8   go to #8

.   .
.   .
.   .

R.17 $Entry1 = 1; Entry2 = 1$; go to #3

.   .
.   .

R.21 $Entry1 = 0; Entry2 = 1$; go to #3

.   .
.   .
.   .

Fig. 1.14. Resolution of procedure controlled nest.

with RANGE:

.

.

for RANGE
for RANGE:

.

.

.

PROGRAM

Fig. 1.15. Sample two-dimensional nest.

given in Fig. 1.14 illustrates this situation. The line numbers and statement labels correspond to those in Fig. 1.13. Thus initial entry to the generation procedure has $Entry1 = 1$ and $Entry2 = 0$, normal deeper nesting has $Entry1 = 0$ and $Entry2 = 0$, normal backtracking has $Entry1 = 0$ and $Entry2 = 1$ (line R.21), and backtracking from the center of the nest has $Entry1 = 1$ and $Entry2 = 1$ (line R.17). A single exit toggle is sufficient since in this resolution the procedure is entered at two distinct places —at line R.4 for moving forwards and at line R.6 for moving backwards.

Compound ranges, constructed from legitimate simple ranges, are sometimes used. The resolution and meaning of the program structures resulting from use of the coincident and sequential specification are pursued in the exercises (and in the next section); the product specification is discussed now.

The form given in Fig. 1.12 only illustrates the nesting of iterations requiring a single indentation. We find extensive application of the nesting of two dimensional iterations (e.g. see Fig. 1.15) and even n-dimensional iterations. The nesting of an n-dimensional iteration is a nest within a nest as illustrated in segment A of Fig. 1.16. This simple nested nest is equivalent to the nest with product range quantifier shown as segment B in Fig. 1.16. As with iterations, however, there are situations in which the product range cannot be used due to the existence of program in one nest but not the other. Also it should be noted that the index of the internal (i.e. second) nest generally depends on the external nest index—see Fig. 7.3.

	"A nested nest"
A.1	with RANGE 1
	.
A.2	.
	.
A.3	with RANGE 2
	.
A.4	.
	.
A.5	ITERATION QUANTIFIER
	.
A.6	.
	.
	(PROGRAM)
	.
A.7	.
	.
A.8	CENTER OF NEST

	"Equivalent product range construction"
B.1	with RANGE 1; RANGE 2
	.
B.2	.
	.
B.3	ITERATION QUANTIFIER
	.
B.4	.
	.
B.5	CENTER OF NEST

FIG. 1.16. Nested nest constructions.

Variably nested nests are occasionally required, but, due to the rarity of such complex constructions, we refrain from notational generalization.

1.5.3. *Auxiliary Transfer of Control Statements*

The statements

reiterate

and

exit from loop

are at times used in **PROGRAM** 2 and **PROGRAM** 4 of a nest (refer to Fig. 1.12). They refer to the level of iteration given by the current value of the with-statement dummy index. They are equivalent to "go to #5" and "go to #9" respectively in the corresponding programs of the resolved form illustrated in Fig. 1.13. Several new transfer of control statements are required: they are

nest deeper

which is equivalent to a "go to #1" in the resolved form,

exit from nest

which is equivalent to a "go to #10" in the resolved form,

backtrack

which is equivalent to a "go to #3" in the resolved form, and

go to center

which is equivalent to a "go to #7" in the resolved form.

1.5.4. *Recursive Operations*

Consider the task of evaluating a determinant by the well-known (but inefficient—see § 2.3.3) technique of expansion by minors. If

$$A_{0 \text{ to } n-1, \, 0 \text{ to } n-1}$$

is an n by n matrix of real numbers, then

$$det(A) = \sum_{j=0 \text{ to } n-1}[(-1)^j \, det(A^{(0,j)}) \times A_{0,j}], \qquad (1.17)$$

where $A^{(0,j)}$ is the $(n-1)$ by $(n-1)$ matrix obtained from A by deletion of row 0 and column j. Using the fact that $det((x)) = x$, where (x) is a 1 by 1 matrix with sole entry x (or $det(\phi) = 1$, where ϕ represents the 0 by 0 matrix), one may calculate $det(A)$ by repeated application of (1.17) to smaller and smaller matrices. From this point of view, determinant evaluation is a summation whose summand is a similar summation, and so forth, the range of each successive nested summation being dependent on the current progress of the preceding summations.

No standard scientific language exists for expressing such recursive operations, although some programming languages (notably Algol) permit "recursive procedures", procedures which may call themselves as part of their evaluation. The chief notational problem is the manner in which the operand–operation dependence is expressed (e.g. how should the information contained in the clause "where $A^{(0,j)}$ is ..." of (1.17) be denoted). The development of concise and readable "recursive terminology" and associated computing techniques will surely receive much attention in the future.

In our language, recursive operations are expressed in terms of the nest notation. A procedure for evaluating a determinant by the method described in (1.17) is given in Fig. 1.17 as segment D. The levels of the nest correspond to the various summations; when i

D.0 $det(A)$

D.1   with $i = n, n-1, \ldots, 1$ as $M = $ (set) $n, n \sim \{j_i\}, \ldots$:

D.2   .

D.3 $S_i = 0$

D.4   for $j_i \in M$:

D.5   .

D.6 $S_0 = 1$

D.7 $S_i + (-1)^{|M \cap j_i|} S_{i-1} \times A_{n-i, j_i} \to S_i$

D.8 $det(A) = S_n$

R.0 $det(A)$

R.1 $det(\ ) = auxdet(0, n, 0)$; exit from procedure

R.T.0 $auxdet(i, M, j)$

R.T.1   (set) M

R.T.2   if $i = n$: $auxdet(\ ) = 1$; exit from procedure

R.T.3   otherwise:

R.T.4 $auxdet(\ ) = \sum_{j \in M} [(-1)^{|M \cap j|} A_{i,j} \; auxdet(i+1, M \sim \{j\}, j)]$

R.T.5   exit from procedure

FIG. 1.17. Recursive determinant evaluation.

equals n we have the summation of (1.17) specifically. The quantities S_i, $i = n, n-1,$ $\ldots, 1$, are the partial sums; the formula "$S_0 = 1$" at line D.6 declares the value of a 0 by 0 determinant to be unity. The actual summation takes place at line D.7; at that time, S_{i-1} is the value of the lower order determinant. The set M marks the columns not yet crossed out for use in selecting minors (line D.4) and in determining the proper signs of the cofactors (line D.7). Note the use of a coincident range specification in the with-clause; M is initially (when $i = n$) equal to $\{0,1,\ldots,n-1\}$ and then loses and gains column indices j_i as the row index i itself decreases and increases, respectively.

For comparison, we have given as segment R in Fig. 1.17 this determinant evaluation as it might be written were recursive procedures used in the language. The notation is essentially the same as in segment D, although due to "stacking" of the arguments of recursive procedures [57], explicit dependence of j on the "level" index i need not be indicated.

EXERCISES

1. State the conditions which must apply to the operands of each of the set-theoretic operations ($\cup$, $\cap$, $\sim$, $\triangle$, $\oplus$, $\ominus$) in order that the operations express a legitimate set-theoretic sequence in a nest quantifier. List the inverses. (For example, union may be used if the sets are disjoint; the inverse is set subtraction.)

2. Resolve segment D of Fig. 1.17 in terms of formulae and transfer of control statements. Note carefully the order in which the incrementing and decrementing of the coincident nest indices i and M must be accomplished.

3. Resolve the program

with RANGE 1, and then RANGE 2:

for RANGE:

in terms of nests with only simple range specifications.

4. Under what conditions are use of the imperatives "backtrack" and "exit from loop" equivalent transfer of control commands?

5. Write a program to calculate all vectors $(d_0, d_1, \ldots, d_{n-1})$ subject to the conditions $0 \leqslant d_0 \leqslant d_1 \leqslant \ldots \leqslant d_{n-1} \leqslant n-1$.

*6. Discuss the relationship between and relative merits of the nest notation and recursive procedures. Consider items such as naturalness, adequacy, usefulness, transparency of resulting programs, implementation, etc. (Actually, this exercise cannot be effectually attacked at this point. An ambitious reader interested in computer science might wish to return to this question after studying Chapters 4 and 5.)

*7. Write a program for evaluating a determinant using the "process of diagonalization", where by appropriate "equivalence operations" (e.g. subtracting a multiple of one column from another column) zeros are obtained for a "triangular" set of $n(n-1)/2$ entries of the matrix yielding the determinant as the product of n "diagonal" entries.

8. Why is use of the statement

reiterate

within **PROGRAM** 3 of a nest (see Fig. 1.12) most likely a mistake? What auxiliary transfer of control statement should be used at the center of a nest to exit from that program (i.e. to continue from the point when **PROGRAM** 3 is completed)?

9. In a nested nest, the possibility of program at the center of the innermost nest just prior to deeper nesting in the outermost nest is suggested in Fig. 1.16 between lines A.6 and A.7. How should the auxiliary transfer of control statements "nest deeper", "exit from nest", "backtrack", and "go to center" probably be interpreted in this **PROGRAM**? Devise a form for nested nests which eliminates this ambiguity.

10. The nest notation resolution of Fig. 1.13 has the (general) iteration quantifier also resolved in terms of formulae and transfer of control statements. Resolve the sample nest notation of Fig. 1.12 using an unresolved for-statement for the iteration quantifier.

11. Write out the with-statement corresponding to the "recursive procedure" resolution (i.e. *with*( ) is a recursive procedure) of the "nest" given in Fig. 1.18.

M.1	execute *with*(1)
M.2	PROGRAM 6
.	
.	
.	

R.0	*with*(k)
R.1	if $k > K$:
R.2	PROGRAM 3; exit from procedure
R.3	for RANGE:
R.4	PROGRAM 2
R.5	execute *with*($k+1$)
R.6	PROGRAM 4
R.7	PROGRAM 5; exit from procudere

FIG. 1.18. Recursive procedure nest resolution.

CHAPTER 2

LANGUAGE IMPLEMENTATION
AND PROGRAM EFFICIENCY

THE language presented in Chapter 1 is machine-independent; that is, notational features of the language make no reference to properties of a particular computer. (By *computer* we mean an electronic digital computer which operates serially from a single instruction sequence.) Nevertheless, most combinatorial programs are indeed written for a particular computer, the one made available to the researcher, and it is desirable to minimize losses of efficiency due to inferior translation of algorithms to the language of that computer. Of course, the precise implementation of a language *is* machine-dependent and we do not intend to inflict on the reader a minute account of a particular implementation. We discuss here only certain general aspects of effective implementation of our combinatorial language and investigate the closely related matter of intelligent and efficient application of the notations presented in Chapter 1. (Most of the content of this chapter is based on the author's experience with the language and compiler Madcap [51], designed and written for the Maniac II computer at Los Alamos.)

Note, particularly, the interplay between language design (and associated compiler writing) and computer design. It is this author's opinion that future efficient combinatorial computing will depend greatly on hardware evolution, especially on the development and integration within general computing systems of special devices for performing certain important jobs extremely fast. In fact, one purpose of this chapter is to assert that natural and convenient programming language and efficient calculation are indeed compatible and to point out places where this compatibility can be increased by reasonable computer design.

The computer scientist will note that the implementation of our combinatorial language is more closely related to that of classic procedural languages than to that of "non-numerical" languages such as Lisp [51]. For example, we rely on sequential rather than "linked" storage of lists. This suggests one more important area of computer science research with regard to combinatorial computing—the use of proven "list processing" and symbol manipulative techniques in the implementation of a high-level procedure-oriented language.

The reader more interested in combinatorial algorithms and less interested in computer science research or the details of computer programming may proceed immediately to Chapter 3.

2.1. Data Representation

The data handled by combinatorial programs take many forms. Actually the choice of form for the data, or, in a broader sense, the selection by the researcher of the computer model (see Chapter 3), determines the structure, hence the efficiency, of a calculation. Here, however, we are concerned with a narrower interpretation of data representation, the representation within the computer, as chosen by the translator, of our basic data-types—real numbers (primarily integers), sets, and arrays of these quantities.

2.1.1 *Single-word Forms*

Digital computers are designed to operate basically upon units of information called *words*. A word consists, most often, of some number of "bits" (i.e. binary digits): 32-, 36-, 48-, 60-, and 64-bit words are common. The precise encoding and significance of bit-information within a word depends upon the particular computer being used and upon the specific machine operations applied to it. Nevertheless, most computers do distinguish between and possess facilities for operating upon the two basic forms of data—arithmetic and logical. Logical words are characterized by the independence of most or all of the bits of the word, while arithmetic words possess built-in "carry propagation" between some of the bits of the word, allowing normal binary representation and manipulation of numbers. Real numbers (more precisely, rational approximations of real numbers) are, of course, represented in an arithmetic word form. For computers which possess more than one such form (e.g. "floating point" and "fixed point", and/or "fixed point" and "index") there is the added burden on the translator of choosing the most efficient form to use for each number and of arranging for conversions between forms in mixed expressions. Although this is not a dfficult task for a compiler (actually many languages shift the responsibility onto the programmer by requiring type declarations such as "real" and "integer"), it is perhaps best handled by machine design. Without debating any of the complex questions of floating point format design, we present here a format which allows efficient implementation of the (single-word) data-type *real*.

Consider the general floating point representation of a number R by an exponent e and fraction f, that is

$$R = (2^{p1})^e \times 2^{-p2}f, \quad -2^{p3} < e < 2^{p3}, \quad -2^{p2} < f < 2^{p2} \tag{2.1}$$

for positive integral machine parameters $p1$, $p2$ and $p3$. (Note that the common case in which the base of the exponent is 2 has $p1 = 1$.) An explicit word form is given in Fig. 2.1. If $p1$ divides $p2$, then integers (and indices) as real numbers which have exponent $e = p2/p1$ have their binary point at the far right-hand end of the word. This location provides maximum capacity for integers and convenient communication with index registers. Thus a computer with facility for rapid $p1$ place shifting permits efficient implementation of real number approximations, integers, and indices as a single data-type. (For a 48-bit machine, a reasonable choice of parameters is $p1 = 8$, $p2 = 40$, $p3 = 4$. With the two sign bits, this allows two data-tag bits; possibilities for their use are mentioned later.)

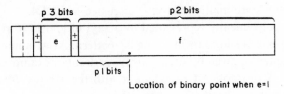

FIG. 2.1. Sample floating point data format.

It has long been accepted that a convenient and efficient representation of sets of natural numbers is by means of a string of bits, the 1-bits of the string giving by their position in the string the elements of the set. For example, labeling the, say, 46 bits of a word (exclusive of the tag bits) from right to left with the numbers 0, 1, ..., 45, the bit pattern

$$10 \ldots \text{(zeros)} \ldots 01011 \qquad (2.2)$$

represents the set $\{0,1,3,45\}$. Such a representation uses the logical form of a computer word since the question of the existence of an element is independent of the existence of other elements of the set.

In this representation, singleton sets become words containing exactly one 1-bit, e.g. 0 ... 0100000 represents $\{5\}$. It is desirable for the computer to have facilities for the insertion, deletion, and test of individual bits of a word. In lieu of this, however, each program should have available an array of singleton sets ($\{0\},\{1\},\{2\}, \ldots$) which may be referenced by simple table look-up from code produced by the compiler. Similarly, it is also useful to have ready access to the array ($\{ \},\{0\},\{0,1\},\{0,1,2\}, \ldots$) = (0,1,2,3, ...) so that statements such as

$$S = \text{(set)} \ 23$$

may be implemented using table look-up.

2.1.2. *Multiple-word Forms*

The range of real numbers and sets allowed by the single-word forms is often not sufficient for combinatorial calculations, and more than one word must be used for expressing a single quantity. Large integers (the chief arithmetic multiple-word form) are quite naturally represented by arrays of single-word forms. For example, the entries of a 1-array A can represent the integer N by having A_0 indicate the number of terms in a polynomial expansion of N with coefficients A_1, A_2, ... That is,

$$N = A_1 \cdot 2^{mp2} + A_2 \cdot 2^{(m-1)p2} + \ \ldots \ + A_m \cdot 2^{p2} + A_{m+1}, \tag{2.3}$$

where $m = A_0$ and $p2$ is the machine parameter discussed in § 2.1.1. An exact rational number with large integral numerator and denominator can, of course, be represented by a pair of multiple-word integers. A rational number representation, particularly useful when most operations are multiplicative, is discussed in § 3.1.4.

Large sets are conveniently represented by arrays of logical words, the entries being considered placed end to end to form a long bit-pattern. As with multiple-word integers, the extent of a set may be given by special array entries. For example, one possible representation of a set S (on a 48-bit machine) is with the 1-array B, where the word B_i, $i = 2$ to $B_1 + 1$, exhibits existence of elements

$$46(B_0 + i - 2) \text{ to } 46(B_0 + i - 1) - 1. \tag{2.4}$$

For sets of natural numbers, B_1 is positive, hence the sign bit of that word can be used to indicate an infinite set, i.e. that all elements larger than $46(B_0 + B_1) - 1$ are in the set. Similarly, the sign bit of B_0 can be used to indicate that all natural numbers less than $46B_0$ are in the set, obviating the storage of many "full" words of bit-pattern for sets which have many consecutive small elements.

Actually, the programming language used in this book makes no distinction between *straightforward* single- and multiple-word calculations. This is done primarily because of the strong machine dependence on such calculations. In practice, the current state of computer and language design demands the use of data-type declarations (e.g. "word-set" or "200 elements", "real" or "80 digits") within a program to assist the compiler in sorting out single- from multiple-word quantities. In anticipation of future design developments, however, we do not use such declarations. Possibly one of the data-tag bits could be used to distinguish between a single-word form and an "absolute address" pointing to a compound (i.e. multiple-word) form. On a computer with interrupt facilities, quantities could even vary in form during a calculation, the interrupt supervisor handling operations on the compound quantities.

2.1.3. *Storage Allocation*

One of the chief advantages of using a programming language is that the compiler arranges for the exact allocation of code and data within the computer memory. Most languages, however, still require that the compiler be told *how many* memory locations will be needed for storage of each array. This is accomplished with a declarative statement such as

$$\text{subscript range: } A_{0 \text{ to } 624}$$

which instructs the compiler to reserve room for 625 entries of the array A. Often in combinatorial computing, the dimensions of an array are either unknown by the programmer (e.g. use of B_i, where $i = 1, 2, \ldots$) or variable (e.g. use of $C_{i,j}$ where $n < i < m$ and $fct(i) \leqslant j \leqslant i$). Thus in our language we do not require dimension declarations.

The implementation of this feature requires allocation of storage for array entries at "execute-time" rather than at "compile-time", hence codes are less efficient than when declarations are required. In most programs, however, this inefficiency is negligible compared to the analysis required in precalculating a range (often tantamount to solving the original problem) or to the loss of storage capacity due to the rectangularization of an irregular array. Of course, a sophisticated language should optionally allow declarations (the statements (1.1) are declarations of a sort).

Storage is allocated for an entry of an undeclared array when the first "store-reference" is made to that entry. This requires a comparison of the current subscript value and the previous minimum and maximum (or perhaps merely maximum) values for that array. If the comparison shows that more space must be allocated, then a supervisory routine must be called to arrange for more room; otherwise, the store can take place "normally". Under the reasonable assumption that storage will have been allocated for an entry before it is used, a "fetch-reference" (or a normal store-reference) to an entry of a 1-array requires one more instruction than when compile-time declarations are used, namely addition of the "base-address" of the array, which is now an execute-time variable. (Most Algol implementations allocate storage for variables declared within a procedure at the time entry to that procedure is made. This type of dynamic storage allocation requires the base-address addition for all variable references. Such addition has been "built-in" to many present-day computers.) It is convenient in this context to view an n-array as a 1-array of $(n-1)$-arrays, and to store as entries in the 1-array the base-addresses of each of the $(n-1)$-arrays. A compound form (see § 2.1.2) is considered an extra dimension of the array (the only dimension for a scalar).

The inefficiency created by the index comparison can be somewhat ameliorated by hardware for rapid inequality comparison between index registers and main memory. A seemingly more significant loss of time occurs when the supervisory routine must allocate new storage space for an entry (or anticipated sequence of entries). Fortunately, however, this time is consumed only when new entries appear, hence is seldom called for in "inner-loop" calculations. For instance, storage for a recursive operation (a nest of loops) will usually have been completely allocated once the center of the nest is reached the first time; this occurs early in the calculation. (A few comments on the "release" of allocated storage are made in § 2.4.2.)

A notation introduced into the language of particular use in utilizing computer memory efficiently is the name convention (see § 1.1.4). When implemented, simple variable names in effect become memory locations; for example, "x" becomes the location in computer storage where the value of the variable x is stored, while "A" (or "$A\_$") becomes essentially the base-address of the array A, the location of entry A_0. A common use of this notation arises when only two rows of the 2-array $B_{i,\_}$ and $B_{i-1,\_}$ need be referenced at each step of a recursion. It is efficient to use two 1-arrays, say $B0$

and $B1$, whose names are interchanged after each step of the recursion, as in

for $i = 0$ to $I-1$, as $B0 = $ INITIAL, $\ldots$, "$B0$" $\leftrightarrow$ "$B1$", $\ldots$:
$$B1 = fct(B0)$$

for example.

As pointed out in § 1.1.4, the name notation is primarily used for transmission of procedure arguments by name rather than by value. This is particularly useful for auxiliary quantities (usually arrays) used in a generation procedure which are not of interest as results from the procedure but which must be preserved and made available for each separate generation using the procedure (the array a of Fig. 5.13, for instance). Of course, when the relevant procedure is referenced from only a single point in the program, it is more efficient not to include such auxiliary variables in the argument list at all. This is possible provided the storage allocated for these variables is not released upon exit from the procedure (see § 2.4.2).

2.1.4. *Toggles and Special Indicators*

There are various features of our language, such as the entry and exit toggles of generation procedures (§ 1.4.2), which require communication of data in bit-size units. Most computers have special facilities (e.g. "sense lights"—program-controlled machine toggles) for accelerated setting and testing of bits of information. We give here a few examples of effective application of such facilities to language implementation.

In the case of the generation procedure toggles, a sense light can be associated with each of the logical variables *Entry* and *Exit* (in practice a single sense light can handle both variables). A statement such as

$$Entry = 0$$

becomes

turn light 13 off

while a test

if $Exit = 1$

becomes

if light 13 is on

for instance. If the names "*Entry*" and "*Exit*" are reserved for this toggle's use only, then the implementation can be accomplished without this machine-dependent terminology appearing in the language.

A sense light can also be used to indicate and delay the execution of a unary operation, such as set complementation, until a more propitious moment. For instance, in executing the iteration

for $a \in A'$:

it is better not to complement the entire set A and then go to the *next*( ) element generation procedure as it may be that the program or computer instruction which scans for ones in the representation of A can equally well scan for zeros. Thus it is generally more efficient to indicate with a single data bit that the complement of the set is to be used.

Another binary indication is concerned with the existence or nonexistence of an element in a set. For instance, how should the notation $(S)_1$ be defined when S is the void set? A simple implementation of this and related notations is to use a sense light to indicate nonexistence. The testing of this light can then be accomplished with machine-independent language such as

$$\text{if nonexistent:} \tag{2.5}$$

Though usable, this implementation and associated language is somewhat awkward, for the existence question can easily occur more than once in the same statement. A more satisfactory solution (again requiring help from computer design) is to isolate from the arithmetic data format a special bit configuration decreed as the "nonexistent number"; call it η. A consistent and useful set of (hardware) operational rules, such as $\eta + x = \eta$, $\{\eta\} = \{\ \}$, etc., can be defined which would, in practice, obviate most tests of the form (2.5). (Actually, for general-purpose computing, numerical as well as combinatorial, two additional special "numbers", ∞ for "infinite" and ? for "indeterminate", should perhaps also be isolated from the data format. In fact, a data-tag might be used to set apart a large number of special numbers, operations on these numbers being handled either by interrupt facilities or explicit hardware.)

EXERCISES

1. Assuming an arithmetic word format as given in Fig. 2.1 with $p1 = 8$, $p2 = 40$, and $p3 = 4$, and a representation of sets as suggested by (2.2), describe computer operations necessary to accomplish $nbr(s)$, where $(s)^1 < 40$, and $set(r)$, for $0 \leqslant r < 2^{40}$.

2. Representing a rational number $P = p_0/p_1$ by the pair (p_0, p_1), where p_0 and p_1 are multiple-word integers, write expressions for the sum, product, and quotient of two rational numbers P and Q.

3. Assuming set operations apply only to single-word bit-patterns, write a program to "make compact" a multiple-word set (represented as indicated by (2.4) and nearby text), i.e. to eliminate words full of 0-bits (or 1-bits) at either end of the bit-pattern.

4. Assuming representation of an n-array as described in the text and using vertical bars for the operation inverse to the "name of" operation (e.g. $|\,"B_{i,j}"\,| \equiv B_{i,j}$), write an expression for $C_{i,j,k}$ in terms of $"C_{0,-,-}"$, i, j, and k. (Hint: $A_n = |\,"A_-"+n|$.)

5. How many sense lights are required to implement the *Entry*1, *Entry*2 and *Exit* toggles of Fig. 1.14?

6. Letting r, η, ∞, and ? represent natural, nonexistent, infinite, and indeterminate numbers respectively, indicate the value of

$$x/0, \ x/\infty, \ x-y, \ x+y, \ |\{x\}|, \ \text{and} \ (x)_1$$

for x and y of each type, e.g. $r + \eta = \eta$.

2.2. Operations

We assume the existence of a basic set of arithmetic and logical word operations permitting straightforward implementation of the operators $+, -, \times, /, ', \cup, \cap, \sim,$ and $\triangle$ used in our language. We discuss here hardware and software implementation of notations which admit less direct treatment.

2.2.1. *Shifting and Exponentiation*

When a set is represented by a string of bits, the $\oplus$ and $\ominus$ operators imply shift operations; for example,

$$0 \ldots 01001101 \oplus \{5\} \quad \text{yields} \quad 0 \ldots 0100110100000,$$
$$0 \ldots 01001101 \ominus \{5\} \quad \text{yields} \quad 0 \ldots 010, \quad \text{and}$$
$$\{5\} \ominus 0 \ldots 01001101 \quad \text{yields} \quad 0 \ldots 0101100.$$

With the right to left representation used here, the first is a left shift, the second is a right shift, and the third is a "reflective" shift in which the order of some of the bits is reversed. The usefulness of these operators (and to some extent their very existence in the language) arises from this efficient implementation. Most computers possess hardware for accomplishing the left and right (single-word) shifting required here. The reflective shift also properly belongs in hardware. It can be realized with simultaneous shifting, in opposite directions, of two "registers" connected at their right (or left) ends. Actually, right and left shifting also require two registers, an extra one to save the "overflow" when the bit-patterns occupy several words. In these cases, however, the normal right end to left end and left end to right end connections are operative.

Since computer multiplication and division instructions are often slow relative to shift and add instructions, it is sometimes better to implement certain multiplicative operations by special program. For instance, on a machine with exponent base two —refer to (2.1) with $p1 = 1$—multiplication and division by a power of 2 is simply an addition in the exponent field. We freely use notation such as

$$\ldots 2^j R \ldots$$

trusting the compiler *not* to form 2^j by an exponentiation routine and then multiply the result by R.

Indeed, the efficient implementation of our language requires a sophisticated general exponentiation process. Program for calculating integral constant powers such as X^5 should be constructed at compile-time from a small number of multiplications (not four—see exercises 9 and 10 of § 3.1). Similarly, the compiler should distinguish between forms such as $X^{p/q}$ (especially when p and q are numerals) and forms such as X^n. In the former case, special information can be transmitted to the exponentiation routine to insure exact computation when the result is integral, while in the latter case, further

analysis must await actual execute-time entry to the routine. The exponentiation procedure itself should analyze for various important special cases: $(-1)^j$, 2^m, X^n with n
integral and small, A^b with b half an integer (provided a square-root instruction exists),
etc. Only as a last resort should a logarithm–multiply–exponential process be adopted—
this case rarely occurs in combinatorial computing.

2.2.2. *Modular Arithmetic*

A common task is to have an index i repeatedly cycle through the least positive residues modulo some natural number n, i.e. to successively assign the numbers 0, 1, 2, ...,
$n-1$, 0, 1, ..., $n-1$, 0, ... to the index i. In our language such a job may be designated
with the notation

$$\text{for } i = 0, \ldots, i+1 \text{ (mod } n) \to i, \ldots : \text{BODY} \tag{2.6}$$

for instance, where the details of the implementation are left unspecified. Such modular
notation may be resolved in terms of simpler statements, but more efficient implementation probably involves computer design. The existence of "counters" in which counting
(and possibly even addition) could be performed modulo a specified number n would
be useful. For small n, a set of circular registers ("ring counters") in which positions of
ones yield numerical values and in which counting corresponds to bit-shifting would be
very efficient in this respect.

A more general solution to this and related problems involves a special integer
divide instruction. For real numbers n and d, $d \neq 0$, imagine a computer instruction
which forms simultaneously

$$q = \left[\frac{n}{d}\right] \quad \text{and} \quad r = n - qd. \tag{2.7}$$

For $d > 0$, r equals n (mod d) as defined in § 1.1.2. This divide instruction can be designed
so that it is especially fast for the common case of n and d integers and $|q|$ small, par-
ticularly so when $|n| < |d|$.

Because of the modular notation, the residue r is probably to be considered the principal result of this special instruction. However, the integral quotient q is useful for
implementation of the "greatest integer" notation, since when $d = 1$, $q = [n]$. Also, in
many combinatorial algorithms (notably base conversions—see § 3.1.1), *both* q and r
are used, hence the define-statement of our language (§ 1.1.1).

2.2.3. *Counting and Locating* 1-*bits*

The use of bit-patterns for the representation of sets calls for bit-manipulative operations other than the basic logical functions. Prominently concerned are implementa-
tions of the notations $|S|$ and $(S)_j$. The cardinality notation entails counting the 1-bits

in a bit-pattern, while the subscript notation involves locating the jth 1-bit in a bit-pattern.

Noting that the set-theoretic intersection of consecutive binary natural numbers differs from the larger by the absence of its rightmost 1-bit (e.g. $110100 \cap 110011 = 110000$), Lehmer [64] suggests a process such as

$$\text{for } b = 0, 1, \ldots \text{ until } W = \{ \ \}:$$
$$W \cap set(nbr(W) - 1) \rightarrow W$$
$$\text{``}|W| = b\text{''}$$

for counting 1-bits in W. Since the number of steps is proportional to the number of 1-bits in W, in the absence of special hardware this process is generally preferable to a direct search exemplified by the "statement"

$$|W| = \sum_{i=0,1,\ldots,L-1 \ni i \in W} (1),$$

where L gives the number of bits to be searched (possibly the logical word length of the computer). Similarly, search for the jth 1-bit in a bit-pattern is illustrated by the program

$$l = -1$$
$$\text{for } k = 1 \text{ to } j:$$
$$\text{if } \exists_{i=l+1,l+2,\ldots,L-1} (i \in W) : l = i$$
$$\text{otherwise: TAKE CARE OF NONEXISTENCE}$$
$$\text{``}l \text{ is the position of the } j\text{th 1-bit''}$$

These searches involve a common task, a scan of the bits of a word with interruption at each 1-bit found. This scan can often be accomplished efficiently by programmed shifting with overflow detection providing the interrupt. Still better, however, is the existence of computer instructions which implement this scan with hardware. Two basic jobs which are easily realized with hardware (see Wells [71]) are to repeatedly shift a register one place, (a) counting the 1-bits which reach a fixed position or (b) counting the steps needed to bring the first 1-bit into a fixed position. With easy access to these counts, single-word implementation of the notations $|S|$ and $(S)_1$ is immediate. Implementation by program of $(S)_j$ and these notations for multiple-word bit-patterns is straightforward (although see Wunderlich [196] for an elaborate implementation useful in certain sieving problems). Implementation of the notation "for $i \in I$:" involving the generation procedure $next (i,I)$—see § 1.4.2—is also straightforward.

Implementation of the notation "if $i \in I$:" is slightly different due to reference to an *arbitrary* bit of the pattern. This test is equivalent to

$$\text{if } (q\text{TH WORD OF } I) \cap \{r\} \neq \{ \ \}:$$

where q and r are defined as the quotient and remainder of i/L as computed by the divide instruction discussed in § 2.2.2 (L is the logical word length of the computer).

2.2.4. *Functions*

Most compiler-systems possess a library of preprogrammed procedures. Functions from the library are automatically incorporated into programs which call them. Thus programs for common mathematical functions such as $log(x)$, *random*, and $\binom{m}{r}$ are written efficiently by specialists and are available for use with no more fuss than required in using addition, multiplication, and square-root extraction. The reader is referred to Lyusternik *et al.* [65] for a discussion of effective approximation algorithms for the elementary functions. We examine here certain aspects of the implementation of more combinatorially oriented functions.

Successive calls to the no-argument function *random* produce a sequence of real numbers arbitrarily selected from the interval 0 to 1. Many methods for the rapid generation of such "pseudo-random" sequences have been proposed [61]. The "multiplicative congruential" method of Lehmer [63] is simple and (for our purposes) reliable and is recommended. Let p be a large prime which just fits in a single word (e.g. if the mantissa of the floating point format has 35 bits, then $p = 2^{35} - 31$ is reasonable—see Hutchinson [60]). Let g be a primitive root [34] of p containing "several" 1-bits in its binary expansion ($g = 5^{13}$ is satisfactory for the above prime) and let x_0 be an arbitrary positive integer less than p. The sequence $x_0, \ldots, x_i = gx_{i-1} \pmod{p}, \ldots, x_{p-2}$ contains the integers 1 to $p-1$ in an order sufficiently mixed up so that the sequence $2^{-c}x_i$, where $c = [log_2(p)]+1$ (e.g. $c = 35$ for the above case), serves nicely as a pseudo-random sequence of words.

Since in practice it is desirable to be able to duplicate a calculation exactly, *random* must act like a generation procedure; it must have an initial entry at which time the sequence is re-started at x_0. Such initialization can be inserted at the beginning of the absolute code by the compiler, and the programmer need only be aware that the initialization occurs when he executes his code from its beginning.

An initialization of another kind is useful with the implementation of the binomial coefficient and factorial notations. While each of these integral functions has simple formulae for its evaluation, it is best to precalculate small tables of exact values which may merely be looked up when requested. The computation of the tables, like the *random* initialization, is inserted by the compiler at the beginning of any code using these notations. For noncompound work, tables of $n!$ for $n = 0$ to (about) 13 and tables of $\binom{n}{r}$ for $n = 0$ to (about) 30, $r = -1$ to $n+1$ (see exercise 8) can be stored. Exact multiplicative calculations involving extremely large factorials and other products are occasionally desired. This subject is pursued from the point of view of a special representation in § 3.1.4 and later exercises.

Functions involving search through a compound quantity being formed as the argument of the function—e.g. "$(S \cup T)_j$" for large sets S and T—invite special implementation. One should not form the union and then go to the precoded routine which locates

the jth one in the bit-pattern, since generally much of the union computation is not needed. The compiler should arrange for the argument calculation to be performed serially,
one word at a time, interleaved with the search for the jth one; when the search is successful, the loop and union calculation is terminated. This implementation applies also to
compound comparisons (e.g. "if $S \subset T \sim V$:"), which are usually implemented as two-
argument functions.

2.2.5. *Control Language*

Implementation of most conditional statements is straightforward. For instance, on a
computer which allows efficient testing for zero and for negative sign, the test "if $a < b$:"
is equivalent to "if $a \neq b$ and sign of $a-b$ is negative". The test "if $a \equiv b$ (mod p):"
is equivalent to "if $a-b$ (mod p) = 0:", while the test "if $S \subset T$:" is equivalent to
"if $S \sim T = \{ \}$:". The important test "if x is even:" can be resolved as "if $0 \in set(x)$:",
this latter test being considerably improved by the existence of a machine test on the last
bit of a word (assuming a format as in Fig. 2.1). The resolution (1.16) illustrates the
implementation of array comparison. The resolutions given in § 1.3.3 answer the question
of compound condition implementation. In using compound conditions, however, one
should note that the order of presentation is sometimes important. For instance,

if CONDITION 1 and (CONDITION 2 or CONDITION 3):

is superior to

if (CONDITION 2 or CONDITION 3) and CONDITION 1:

when CONDITION 1 is usually false. (In this respect see Riesel [67] and Slagle [68].)

Similarly, the resolutions presented in § 1.4 deal with most iteration implementation
questions. Several of the range specifications require library generation procedures. The
for-clause "for $a \subset A$:", for instance, requires a procedure to generate bit-patterns
"covered" by the A bit-pattern (see § 5.1). The procedure placed in the code by the compiler as a result of use of the notation "for i:" receives for an argument the location of a
block of data describing the first array referenced in the body of the loop. This data,
which may vary within the loop itself, is used by the procedure to determine values to
assign to the index i.

The resolutions described in § 1.5 indicate the basic implementation of with-statements. As indicated in exercise 2 of that section, one must be somewhat careful regarding
the order of the statements of the resolution.

Of course, our use of nests allows recursive processes to be expressed in terms of
iterations, hence we do not require an implementation of recursive procedures, i.e.
procedures which may reference themselves during their execution. The reader may
consult one of the numerous descriptions of Algol implementations (Dijkstra [57], for

instance) for a discussion of the handling of recursive procedures. Also, see Floyd [107] and Johansen [115] for a rather different approach to language and implementation of backtracking. (Also, see exercise 11 of § 1.5.)

EXERCISES

1. Using the multiple-word representation given by (2.4) and assuming the set operations used in your program apply only to single-word bit-patterns, write programs to implement

$$S', S \cup T, \text{ and } S \cap T$$

for general sets S and T.

2. Assuming that the operation $2^j R$ is accomplished (efficiently) without multiplication, write a program to form $n \times R$, for n a natural number, which does not use multiplication.

3. Resolve (2.6) in terms of formulae and transfer of control statements.

*4. Assuming the existence of the $log(x)$ and $exp(x)$ functions, write a general (single-word) exponentiation function of at least two arguments which incorporates handling of the special cases mentioned in the text. Use the arithmetic format of Fig. 2.1 with $p1 = 8, p2 = 40, p3 = 4$.

5. For integral n and d, $d > 0$, write a procedure which yields q and r as defined by (2.7) without using division or the "greatest integer" notation.

6. Write a $next(i,B)$ generation procedure, where the set B is represented as suggested by (2.4). Assume that all set operations, including $(S)_1$, apply only to single-word bit-patterns capable of representing subsets of $\{0,1,\ldots,45\}$.

7. Assuming the 48-bit arithmetic and logical word format discussed in § 2.1.1 and the set representation given by (2.4), write a program which generates random subsets of $\{0,1,\ldots,919\}$.

8. Given a table of $C_{n,r} = \binom{n}{r}$ for $n = 2$ to 30 and $r = 1$ to $[n/2]$, write a function to calculate $\binom{n}{r}$ for $n = 0$ to 30, $r = -1$ to $n+1$.

9. Write a program to calculate a table of cubes, $Q_i = i^3$ for $i = 1$ to 1000, without using multiplication. (Hint: Use the binomial expansion [3].)

10. Translate the following program into a (probably less efficient) program involving only real number operations which accomplishes the same job [69].

$$s = \{0,1\}$$

for $i = 1, 2, \ldots$, until $s \cap i = \{ \}$:

$$set[nbr(s \oplus \{1\}) + nbr(s)] \rightarrow s$$

EUREKA!

11. If a and b are positive integers, the expression $a = bq + r$, $0 \leqslant r < b$, uniquely defines an integral quotient q and integral remainder r for the operation a/b. The binary representation of q may be calculated by a division algorithm which finds the largest power of 2, say 2^j, such that $2^j b \leqslant a$, then the largest power of 2, say 2^k, such that $2^k b \leqslant a - 2^j b$, etc. Write a program which calculates sets Q and R such that $nbr(A) = nbr(B) nbr(Q) + nbr(R)$, $0 \leqslant nbr(R) < nbr(B)$, for given sets A and B.

2.3. Program Optimization

The efficiency of a calculation depends upon both the effectiveness of the computational algorithm and the painstaking of its realization as a computer program. This section is primarily concerned with the second of these considerations, with optimization of program writing, although, often, the two aspects of efficiency cannot be separated due to inadequacy of algorithm specification or concurrent development of program and algorithm.

Of course, without a detailed knowledge of the workings of the available computer and compiler, one is always limited in the optimization he can achieve. Thus we discuss here only certain "commonsense" aspects of efficient programming. The scientist must be prepared to experiment with both methods and programming whenever a long calculation is in the offing.

2.3.1. *Precomputation*

An important principle of efficient computing is that a computation should be carried out as soon as all the quantities required in that computation are available. Thus, for example, in computing direction cosines, the quantity

$$D = \sqrt{(A^2 + B^2 + C^2)}$$

should be precomputed once and for all as soon as the values of A, B, and C are known, and then this value of D used in evaluating

$$c_1 = A/D; \quad c_2 = B/D; \quad c_3 = C/D.$$

Similarly, quantities not dependent upon an inner loop index should be calculated external to the loop controlled by that index. Consequently, programs such as

$$\text{for } i \in I:$$
$$f_i = \text{EXPRESSION}$$
$$\text{for } j \in J:$$
$$g_{i,j} = FUNCTION(j, f_i)$$

occur frequently.

When applied to nested iterations, this principle of precomputation affects even the structure of a calculation and can have significant effects on its efficiency (see § 4.2). In fact, the principle of precomputation is not entirely notational. The concept of computing ahead of time certain quantities which would have to be generated many times within a nest can supply important inspiration to a combinatorial researcher. It was essentially the application of this idea which led to Parker's breakthrough in computing orthogonal Latin squares (see § 7.4).

While precomputation often shortens the duration of a calculation, it usually demands more storage space (for saving the precomputed quantities). Also, the resulting programs are more detailed and complex (further removed from natural straightforward language), hence the chance for error is increased. Some of these shortcomings can be mitigated by use of an elaborate compiler which automatically performs optimizations such as the precomputation of common expressions and the removal from a loop of index-independent calculations as illustrated above. For the most part, however, one must seek a happy medium between storage and time and between naturalness and efficiency. As a word of warning, we point out that when using a precomputed quantity, one must be extremely careful to check that no components of that quantity have changed in value since its computation.

Since our language allows components of a range specification to vary during the execution of an iteration or nest, the burden of precomputing fixed values for these parameters rests with the programmer. Thus pairs of statements such as

$$I = \text{EXPRESSION}$$
$$\text{for } i = 1 \text{ to } I:$$

occur often in our programs. Relative to the warning of the previous paragraph, such precomputation is not possible when the iteration is part of a nest and the upper limit depends on the next index—unless, of course, that dependency is explicitly expressed with a subscript on the precomputed upper limit. For instance, the program

$$\text{with } j \in J$$
$$\ddots$$
$$I_j = 2A_j + B_j$$
$$\text{for } i_j = 1 \text{ to } I_j$$
$$\ddots$$

is correct, but it would be incorrect to use an unscripted I for the precomputed upper limit.

2.3.2. *Specialization Versus Generalization*

General programs, programs written to handle many cases, are normally less efficient than special programs tailored to deal with a particular situation only. For instance, in many contexts, reference to a variable n whose value is not fixed until execute-time requires more code than reference to a constant fixed at compile-time. Some languages in fact possess facilities enabling a programmer to distinguish between compile-time "parameters" and execute-time "variables". (A truly comprehensive compiler—with attendant loss of compiling speed of course—could make many such distinctions without explicit hints from the programmer.) The merit of such distinction depends largely on the problem itself; it ranges from significant for a long-running problem to negligible (or even negative) for an experimental, ever-changing, short-running problem.

Multiple-word calculations are, of course, significantly slower than single-word calculations. Thus the use of compound forms for the sake of generality should be avoided whenever possible. Our language makes no distinction between such calculations, trusting proper compiler and computer design with the problem (see § 2.1.2). When this is not possible, and when data-type declarations are employed to help the compiler distinguish between single- and multiple-word quantities, we urge that the burden be placed on multiple-word calculation; that is, a set should be considered a one-word bit-pattern unless otherwise specified.

Questions of specialization versus generalization also are not solely notational. In experimental combinatorial computing it is often best not to program for every possible contingency but to adopt a "wait and see" policy, generalizing the program only when it becomes clear that certain special cases do indeed arise. In this way, extraneous work can many times be avoided. Also, such a policy often leads to effective interaction between human and electronic computing, since dealing with a particular case personally is sometimes easier than programming for its automatic surveillance.

2.3.3. *Use of Sophisticated Notations*

Many of the features of our language are several levels of resolution removed from the absolute machine code that actually specifies and controls the calculation. Thus, although convenient, their use can conceal inefficiencies. For instance, the statement

$$P = \bigcup_{i\,=\,3,5,\ldots,9999} \ni \forall_{j\,=\,3,5,\ldots,\sqrt{(i)}} \; [i \not\equiv 0 (\bmod j)] \; \{i\}$$

albeit straightforward and correct, is indeed a poor way to form the set of odd primes less than 9999. It is instructive to look at the ways in which such notation can imply inefficient computation.

Note first that unless the compiler detects that i does not change in the j loop, the quantity $\sqrt{(i)}$ will be recomputed for each value of j. Second, the tests for divisibility are accomplished with division instructions, whereas by using a less direct sieve method (see § 6.1) only additions are required. Finally, this notation makes no provision for using partially gathered information. For instance, the division $317/15$ will be performed even though 15 is already known to be composite.

The last of these is both the most insidious and the most important. It is easy when using convenient notation, especially when programming recursively, to lose sight of computations which are needlessly being repeated. Of course, the blame should not be placed entirely on the notation as the closely related question of algorithm design is certainly involved. A good example is the determinant evaluation of Fig. 1.17. In that recursive summation many cofactors are being formed more than once, for instance, $A_{n-2,n-2}A_{n-1,n-1} - A_{n-2,n-1}A_{n-1,n-2}$ is formed $(n-2)!$ times, once for each term $A_{0,j_0}A_{1,j_1} \cdots A_{n-3,j_{n-3}}$, where $(j_0,j_1,\ldots,j_{n-3})$ is a permutation of $(0,1,\ldots,n-3)$. This evaluation requires of the order of $n!$ operations whereas a scheme which eliminates such repeated computation (see § 4.4.2) requires only about $n \times 2^{n-1}$ operations. (Of course, for determinant evaluation itself still faster methods are available—see exercise 7 of § 1.5.)

Certainly we do not intend to imply that sophisticated notations are of little use; we suggest only that they should not be used indiscriminately. For a large number of calculations, top efficiency is of much less importance than readability and swift attainment of results. For the most part, only the inner-loops and heavily traveled paths of long running calculations need be finely tailored for maximum efficiency. One would hope that sophistication of compiler writing itself will eventually suppress most optimization worries from the ken of the programmer.

EXERCISES

1. The program

   for $n = 1$ to 15; $k = 1$ to n:

       if $k = 1$ or $k = n$: $S_{n,k} = 1$

       otherwise: $S_{n,k} = S_{n-1,k-1} + kS_{n-1,k}$

calculates a table of *Stirling numbers of the second kind* (see Appendix I). Write a more efficient recursion and compare the "running times" of the two programs (perhaps only by "desk analysis").

2. Compare the running times of segments A and B of Fig. 1.11. Use $k = 3$ in segment B, of course.

3. Assuming P is the set of primes less than 10000, write a one-line test of the form

$$\text{if } \exists_{\text{RANGE}} (\forall_{\text{RANGE}}(\text{CONDITION})):$$

which would be true if a counterexample to Goldbach's conjecture (every even number is the sum of two primes) existed for some even number < 10002. (Compare for efficiency with the program given in Fig. 6.4.)

4. Discuss the topic "Storage versus time in combinatorial computing".

2.4. System Organization—Procedures

Compiling (i.e. translating from source language to absolute machine code) is best viewed within the framework of a "software system", a class of service programs (e.g. compilers, correctors, supervisors, library procedures) designed to assist the user in editing and executing his problem programs. We are concerned here with the influence of system operation upon the language and its implementation. This influence is primarily felt with regard to program segmentation, procedures, and associated argument handling.

2.4.1. *Program Segmentation*

A program often consists of a number of more or less self-contained pieces, each piece accomplishing a particular job, with communication between pieces via a few common variables. A piece itself may logically consist of smaller jobs of equivalent stature within that piece, but subsidiary to the program as a whole. Thus a program may quite naturally have a tree-like structure, the nodes of the "tree" (§ 3.4.4) corresponding to program segments. For practical reasons also—to facilitate correction and recompilation and to enable external reference such as required by "conversational debugging"—it is convenient to have a large program subdivided into smaller, more manageable units. These conceptual and practical demands on program segmentation are not incompatible, hence it is advantageous to adopt a common system, as sketched in § 1.1.1, for both uses.

Since most algorithms presented in this book are relatively compact, we utilize only a fraction of the potential of program segmentation. In fact our exploitation here is chiefly expository, using segment labels and line numbers for textual reference and for visual delimitation of independent programs and procedures. Actually, most algorithms are given here as procedures, with communication via explicitly listed arguments. While convenient, this is often inefficient as transmission of arguments can be time-consuming. Of course, in cases where argument transmission is especially onerous, a procedure code can be copied "in-line" by the programmer with appropriate name substitution so that no "linkage" at all is required.

The following rules for determining the scope of names in a segmented program allow efficient communication between segments and are used in this book: sibling segments, that is, segments with the same segmentation tree ancestry (e.g. A.B.B and A.B.C), are considered *co-segments*; all names used within a set of co-segments are in common. Names for variables which may be fetched before being (re)stored when the set of co-segments is executed are in common with the same names used in the father segment. Names for variables which are always stored before being fetched are considered "local" names and are not equated to similar names in the father segment.

These rules (which can be extended to procedure names and to statement labels) require "two-pass" compilation for their implementation, but permit natural program subdivision with little or no explicit declaration of the names used within a segment. [The two-pass compilation is also needed to distinguish "$f(x+1)$" from "$f \times (x+1)$", to establish data-types, and, in general, to gather information for the translation of context-dependent language.]

Note that a single procedure nested within a segment has private names for its internal variables, yet may communicate with its father (also without declarations) via input variables not listed as arguments. Furthermore, since co-segments are name-wise equivalent, the system itself can effect limited segmenting of a practical nature for naïve programmers.

2.4.2. *Procedure Linkage*

A procedure is a segment which has been given a name for reference via a mnemonic-function-call notation. An implementation of procedure notation is sketched in Fig. 2.2. This illustrates the primary types of linkage that exist between a procedure and its reference: transfer of control, transmission and assignment of argument values, and transmission of output results.

Note, with regard to argument transmission, that because an arbitrary expression may appear as a value argument in the reference, value-transmission is often tantamount to precomputation. (Name-transmission may also be extended to allow arbitrary expressions as arguments. The implementation of this generalized notation is more difficult than sketched in § 2.1.3, however, and we do not use such notation in this book.) The chief use of name-transmission is to simplify communication of array data. In Fig.

"Using procedure notation"

A.6 $\qquad y = bc + proc(x-1, "A")/g$

A.B.0 $\quad proc(z, "M")$

A.B.1 $\qquad$ PROGRAM INVOLVING z and M.

A.B.13 $\qquad proc(\;) = $ EXPRESSION

"Without using procedure notation"

A.6 $\qquad$ go to #2

#1 $\quad y = bc + d/g$

A.B.0 #2 $\quad z = x-1;\; "M" = "A";$ ARRANGE FOR EXIT TO #1

A.B.1 $\qquad$ PROGRAM INVOLVING z and M.

A.B.13 $\qquad d = $ EXPRESSION $\qquad$ "d is a dummy"

$\qquad\qquad$ go to (#1)

FIG. 2.2. Linkage between a procedure and its reference.

2.2, for instance, if A is an array, then references to M_k in the procedure use A_k from its original location. Name-transmission also permits output of auxiliary results; the appearance of M on the left side of a formula within the procedure produces a change in the value of A external to the procedure.

Although shown in the figure as a dummy variable, the output result d—the value of the procedure—is usually transmitted via the computer "accumulator". Also, the transfer of control statement "go to #2" in reality interrupts the execution of the formula on line

A.6. All "registers" are preserved and when return from the procedure occurs, execution is continued from the precise point at which the formula was interrupted (division by g and addition of the already computed term bc remains to be done).

In this example, the procedure is nested within the program which used it. Thus all names for "stored-before-fetched" variables within the procedure are local names: they are to be distinguished from identical names external to the procedure. Storage allocated for these variables is therefore not used outside of the procedure (strictly speaking, outside of its set of co-procedures), hence may be *released* for other use. Within a system using dynamic storage allocation (§ 2.1.3), this is accomplished by a supervisory routine called immediately prior to exit from the procedure. If declarations are used, then a supervisory routine is also called upon entry to the procedure in order to allocate storage as declared.

2.4.3. *Compilation and Execution*

The process of compilation is in effect a precomputation, a once-and-for-all analysis and translation of a source program into a form executable by a computer and its associated supervisory routines. Compilation and supervision are complementary processes whose "sum" measures the sophistication of the source language. In general, time efficiency is proportional to the percentage of compilation achieved. For instance, a system with no compilation and 100% supervision—an "interpretive" system—produces extremely slow-running executable codes (the source program itself!). On the other hand, systems with no supervision (compilation produces a finished absolute code) require excessively lengthy compilation for a reasonably sophisticated and flexible source language. A compromise system weighted in favor of time efficiency of executable code is perhaps best for general combinatorial computing.

In such a compiler-oriented system, the compilation time itself can be significant, especially when the compiler is asked to perform many time-consuming optimizations. This is particularly annoying for experimental computing whichentails frequent source program alteration. Several possibilities for bettering this situation exist.

First, and least attractive, is to introduce "compiler controlling" vocabulary into the source language. That is, declarations and special language such as "n is 100" (to define n as a compile-time parameter which will not change at execute-time) can be used to help the compiler avoid time-consuming program analysis in producing an efficient absolute code. Another possibility is to make available two (or more) compilers of different sophistication. One performs speedy translation but produces inefficient executable code for use while a program is unsettled; the other produces a polished executable code at the expense of compilation time. A third and quite natural approach is to render recompilation lss onerous by use of a system which allows compilation of a program in pieces, the pieces being merged at time of execution. In the actual use of our combinatorial language, it is convenient to treat a segment or set of co-segments as an independent piece of program which may be compiled, "debugged", altered, and recompiled without affecting other parts of the program.

Perhaps, in practice, an approach combining features of all three possibilities is best. The point is that there do exist ways of obtaining efficient executable codes (a basic goal of combinatorial computing) without undue sacrifice of language elegance and convenient communication between scientist and computer.

EXERCISES

1. A name for a variable is considered local to a segment and its co-segments if the variable is always given a value before being used within the set of co-segments. Define appropriate rules for the interpretation of procedure names and statement labels. Construct your definitions so that library procedures are handled naturally.

2. Argument names appearing in a procedure heading are considered local to that procedure and its co-procedures. Is this consistent with our segmentation interpretation?

3. Discuss and sketch implementation for a notation using quotation marks around a general expression (rather than just a name). For instance, how should

$$\text{``}D\text{''} = \text{``}\sqrt{(A^2 + B^2 + C^2)}\text{''}$$

PROGRAM

$$c_1 = A/D; \quad c_2 = B/D; \quad c_3 = C/d$$

be handled? In what way would this program differ from

$$D = \sqrt{(A^2 + B^2 + C^2)}$$

PROGRAM

$$c_1 = A/D; \quad c_2 = B/D; \quad c_3 = C/D.$$

CHAPTER 3

COMPUTER REPRESENTATION
OF MATHEMATICAL OBJECTS

FROM a "physical" problem which a computer is to solve one must first erect a model —a representation of pertinent entities—in a form comprehensible to the computer. This is often a two-step process: first a mathematical model for the physical problem (e.g. a linear graph as a model for a city street plan) is derived and then a computer representation for the mathematical object (e.g. a 2-array of zeros and ones as a representation for the graph) is constructed. We are primarily concerned here with the second step, with constructing computer representations of mathematical objects. In this respect, this chapter can be considered an extension of § 2.1. It serves as an introduction to the applications appearing in this book and in the literature. Also, for those less familiar with combinatorial mathematics, it may serve as an introduction to basic combinatorial terminology.

It should be noted that the proper choice of a model depends heavily upon the individual computer system as well as upon particular features of an intended study. Also, generally speaking, the more compact the representation, the more difficult it is to extract the usable information—a happy medium should be found for each problem. Thus we often present alternative representations. However, as discussed in Chapter 2, our representations (and resulting algorithms) are indeed biased toward systems with convenient bit-manipulative facilities.

3.1. Natural Numbers

Basic to combinatorial computing are the positive and negative integers and zero. In many number-theoretic investigations these numbers themselves are the primary objects of study. In other combinatorial studies these numbers serve either to label or to count various discrete objects under consideration. Actually, the negative integers are seldom needed—the natural numbers 0, 1, 2, ... serve nicely as both *indices* (labels) and *tallies* (counts). This section discusses the common and useful representations for the natural numbers.

We should point out that for clarity of textual presentation it is often necessary to use nonnumeric labels such as $a, b, c, \ldots$, or $v_1, v_2, v_3 \ldots$. However, this does not imply their use within the computer. In fact, our language and algorithms rely significantly on the representation of labels as natural numbers.

3.1.1. *Positional Representation*

The most important representation of a natural number is by means of the conventional *positional* notation. For a fixed integer B, $B \geqslant 2$, the *base* or *radix*, and arbitrarily large integer n, each natural number N less than B^n may be represented by an ordered set of "digits" $d_0, d_1, \ldots, d_{n-1}$ defined by the expansion

$$N = d_0 + d_1 B + d_2 B^2 + \ldots d_{n-1} B^{n-1}, \tag{3.1}$$

where each d_i, $i = 0$ to $n-1$, satisfies the inequality

$$0 \leqslant d_i < B$$

to insure that the representation is unique. For $B = 10$ (writing the digits in order $d_{n-1} d_{n-2} \ldots d_0$ and omitting leftmost zeros) this is the conventional decimal notation. For $B = 2$ this is the common binary system of representation used by most high-speed digital computers.

It is occasionally advantageous to express numbers relative to a base different from that used in the accordant hardware representation—the *standard representation*. Such representations may be formed "from scratch" using programmed operations (see §§ 4.1 and 5.1) or may be derived by conversion from the existing numbers in standard representation. The conversion may be accomplished from either end, i.e. the new digits may be calculated in order $d_0, d_1, \ldots, d_{n-1}$, or in order $d_{n-1}, d_{n-2}, \ldots, d_0$. The former is achieved by successively dividing first N, and then each resulting quotient, by B; the resulting remainders provide the digits. That is, $q_0 = N$, $q_i = d_i + q_{i+1} B$ for $i = 0$ to $n-1$. The latter sequence is formed by dividing N by B^{n-1}, the resulting remainder by B^{n-2}, and so forth; the resulting quotients provide the digits. That is, $r_{n-1} = N$, $r_i = r_{i-1} + d_i B$ for $i = n-1, n-2, \ldots, 0$. Computing the digits in ascending order of subscript is slightly more efficient as the process may be terminated when a zero quotient appears, while for the descending process, since we begin with a sufficiently large n so that $N < B^n$, several zero digits may be inadvertently computed.

This descending conversion, however, which is given in Fig. 3.1, is perhaps more useful in representations closely related to the positional model just discussed. An

C.1 $r = N$

C.2   for $i = n-1, n-2, \ldots, 1$:

C.3   define d_i, r by $\dfrac{r}{B^i}$ "r (old) $= r$ (new) $+ d_i B^i$"

C.4 $d_0 = r$

FIG. 3.1. (Descending) conversion to radix B.

important representation of this type is the so-called *factorial representation* where each N, $N < n!$ (n an arbitrary positive integer) is given by an ordered set of *factorial digits* $a_1, a_2, \ldots, a_{n-1}$ defined by the expansion

$$N = a_1 + a_2 \cdot 2! + \ldots + a_{n-1}(n-1)!, \tag{3.2}$$

where each a_i, $i = 1$ to $n-1$ ($a_0 = 0$ is sometimes introduced) satisfies the inequality

$$0 \leqslant a_i \leqslant i \tag{3.3}$$

to insure that the representation is unique. Only minor modifications of the conversion of Fig. 3.1—replacing B^i of line C.3 by $i!$ and shortening the iteration by one—are needed to enable it to calculate the factorial digits associated with a given natural number N (see § 5.2).

3.1.2. *Gray Codes*

A *Gray code* [162] is a function of the numbers $0, 1, \ldots, 2^{n-1}$ onto themselves (i.e. a permutation) such that the images of successive numbers differ at a single-bit position in their binary representation. For example, 0, 1, 3, 2 (or in binary 00, 01, 11, 10) is a Gray code for 0, 1, 2, 3. For arbitrary n, many distinct Gray codes exist (in fact, the exact number is unknown for $n \geqslant 5$—see § 4.2.4). Given a Gray code, the *Gray code representation* of N is the image of N under the code.

A particular Gray code, of interest because of its simplicity, is defined as follows: Let n be an arbitrary positive integer. Let $b_0, b_1, \ldots, b_{n-1}$ be the binary digits of a natural number N, $N < 2^n$, in the standard positional representation. Consider the representation $g_0, g_1, \ldots, g_{n-1}$, where $g_j \in \{0,1\}$ and

$$g_j = b_j \, \Delta b_{j+1}, \, j = 0 \text{ to } n-1, \, b_n = 0. \tag{3.4}$$

(Symmetric subtraction of bits is mentioned in § 1.2.2.) For this to be a Gray code, the representation of $g_0^*, g_1^*, \ldots, g_{n-1}^*$ for $N+1$ must differ from $g_0, g_1, \ldots, g_{n-1}$ in precisely one place, i.e. there must exist a k such that $g_k^* \neq g_k$, and $g_i^* = g_i$ for $i \neq k$. Indeed, let k be the smallest index for which $b_k = 0$ (i.e. when one is added to N, "carry-propagation" reaches just to position k). Now bits $b_0^*, b_1^*, \ldots, b_k^*$ of $N+1$ are complements of $b_0, b_1, \ldots, b_k$ and thus $g_j^* = g_j$ for $j = 0$ to $k-1$. Since $b_{k+1}^* = b_{k+1}$ and $b_k^* \neq b_k$, we have $g_k^* \neq g_k$. Furthermore, since none of the bits beyond b_k were changed by addition of one to N (i.e. $b_j^* = b_j$ for $j = k+1$ to $n-1$), we have $g_j^* = g_j$ for $j = k+1$ to $n-1$. Therefore, (3.4) does define a Gray code. Note further that g_k^* equals 0 or 1 according as b_{k+1} equals 1 or 0 respectively.

To express N in conventional binary given its Gray code representation $g_0, g_1, \ldots, g_{n-1}$ as defined above, the inverse of (3.4) may be employed, namely

$$b_j = b_{j+1} \, \Delta g_j, \, j = n-1 \text{ to } 0, \, b_n = 0. \tag{3.5}$$

Further discussion of this particular Gray code appears in § 5.1. The generation of other Gray codes is treated in § 4.2.4.

3.1.3. *Residue Representation*

Let $m_1, m_2, \ldots, m_n$ be n positive integers relatively prime in pairs and let $M = \prod_{j=1 \text{ to } n}(m_j)$. That each number N less than M is uniquely determined by its n least positive residues modulo m_j, $j = 1$ to n, is a well-known number-theoretic result. The representation of N by these residues $r_1, r_2, \ldots, r_n$ is called the *residue* (or *Chinese* or *modular*) *representation*.

If N is represented in this fashion by the 1-array $(r_1, r_2, \ldots, r_n)$ and N^* by $(r_1^*, r_2^*, \ldots, r_n^*)$, then it follows from the elementary rules of modular arithmetic that $N + N^*$ is represented by

$$(r_1 + r_1^* \ (\text{mod } m_1), \ldots, r_n + r_n^* \ (\text{mod } m_n))$$

and NN^* is represented by

$$(r_1 r_1^* \ (\text{mod } m_1), \ldots, r_n r_n^* \ (\text{mod } m_n)).$$

The independence of the r's in these calculations has aroused some interest among computer designers although no actual hardware implementation seems to have appeared (but see Szabó and Tanaka [97]).

The conversion from residue to standard representation, often required in number-theoretic calculations, may be accomplished by means of the venerable *Chinese remainder theorem* [34]. This says that

$$N = \sum_{i=1 \text{ to } n} \left(r_i b_i \frac{M}{m_i} \right) (\text{mod } M), \tag{3.6}$$

where b_i is some integer such that $b_i(M/m_i) - 1$ is a multiple of m_i (i.e. $b_i(M/m_i) \equiv 1 \ (\text{mod } m_i)$). The existence of each b_i is insured by the pairwise relative primeness of the m's which makes m_i and M/m_i relatively prime. A naïve, yet often satisfactory, method of determining b_i is simply to try each of the values $1, 2, \ldots, m_i - 1$ until the correct one is found. For example, the function

      B.0   *inverse(c,m)*

      B.1   *inverse( )* = [first $j = 1$ to $m-1 \ni jc \equiv 1 \ (\text{mod } m)$]

would produce b_i when applied to M/m_i and m_i as arguments. More sophisticated means of determining b_i are pursued in the exercises. In any event, given n, the vectors $m_{1 \text{ to } n}$ and $r_{1 \text{ to } n}$, and a procedure, say *inverse( )*, for calculating each b_i, the formula (3.6) may be evaluated directly—by use of the program

$$M = \prod_{j=1 \text{ to } n}(m_j)$$

$$N = \sum_{j=1 \text{ to } n} \left(r_i \frac{M}{m_i} \ inverse \left(\frac{M}{m_i}, m_i \right) \right) (\text{mod } M) \tag{3.7}$$

for instance.

Additional discussion of this conversion and related topics appears in the exercises and in § 6.1.

3.1.4. *Prime Factor Representation*

In the *prime factor representation* a number is given by the vector of exponents appearing in its prime factorization: one component position in the vector is reserved for each possible prime that can occur. For instance, $140 = 2^2 \times 5 \times 7$ is given by (2,0, 1,1,0). By allowing components of the vector to be negative, rational numbers may also be represented, for instance, $140/121$ is given by $(2,0,1,1,-2,0,\ldots)$. This model is useful for the exact representation of rational numbers in which neither numerator nor denominator has very large prime factors [95]. Multiplicative manipulations with factorials, binomial coefficients, and other products of small integers lend themselves readily to this representation since multiplication reduces to addition (and division to subtraction) of vectors.

Perhaps the primary advantage of this representation is that numerator and denominator of a rational number appear in factored form *without common factors*. Certain algebraic and number-theoretic manipulations such as testing for "square-freeness" and generating factors of a rational number are simplified when this representation is used. Also, the automatic elimination of common factors yields compact and exact results directly without the extra application of the Euclidean algorithm to the monstrous integers obtained as numerator and denominator of the rational number when standard exact multiple-word techniques are applied (see exercise 2 of § 2.1).

Additive operations on numbers in this form involve conversion to and from standard representation, which in turn involves the factorization of integers. This can be quite time-consuming; hence, the prime factor representation is to be avoided when many additive operations are necessary. Also, when very large prime factors occur, the representation should be modified so that only primes appearing in the factorization are explicitly recorded; then, zero components of the vector need not be saved.

3.1.5. *Bit Parity of Multiples of Primes*

We conclude this section on natural numbers with an interesting combinatorial investigation which originated from an empirical study of the binary representation of the prime numbers. In attempting to explain the apparent preponderance of primes which have an odd number of 1-bits in their binary expansion, it was noted that multiples of 3 tend to have an even number of 1-bits (even parity). In fact, the excess number of multiples of 3 less than 2^z with even parity is $3^{(z-1)/2}$ or $2 \times 3^{z/2-1}$ according as n is odd or even. Similar results for some odd primes other than 3 were established after inspection of the results of certain numerical experiments. Simple formulae seem to exist for primes which have 2 as a primitive root and for primes p which have 2 belonging [34] to the exponent $(p-1)/2$ with $(p-1)/2$ even. The known formulae are summarized in Wells [100]. (Note: As can be seen in Appendix II, not all primes produce an excess of even parity multiples.)

"Enter with p and Zp given"

R.1 $r_0 = 1; r_1 = 2$

R.2   for $t = 1, 2, \ldots$, until $r_t = 1$: $2r_t \pmod p \rightarrow r_{t+1}$

R.3 $m =$ first $m = 0, 1, \ldots \ni 2^{m+1} > p$

R.4   for $i = 0$ to $2^m - 1$:

R.5   if $|set(i)|$ is even: $d_{m,i} = 1$; otherwise: $d_{m,i} = -1$

R.6   for $i = 2^m$ to $p-1$: $d_{m,i} = 0$

R.7   for $z = m, m+1, \ldots, Zp$:

R.8   for $i = 0$ to $p-1$:

R.9 $y = i - r_{z(\mathrm{mod}\ t)} \pmod p$

R.10 $d_{z+1,i} = d_{z,i} - d_{z,y}$

R.11   "Calculation of $d_{z,i}$, $z = m$ to Zp, $i = 0$ to $p-1$, is complete"

FIG. 3.2. Calculation of bit parity function for multiples of primes.

The program used in the numerical experiments is an instructive example of a combinatorial algorithm; it is presented in Fig. 3.2. The program is a recursion of $d_{z,i}$, the number of nonnegative integers of the form $np+i$, where $np+i < 2^z$, with even parity minus those with odd parity, i.e.

$$d_{z,i} = \sum_{j = i, i+p, \ldots,\ \text{until } j\ >\ 2^z} \left((-1)^{|set(j)|} \right). \tag{3.8}$$

One notes that each binary number j, $j < 2^z$, has the opposite parity of the binary number $j+2^z$, thus

$$d_{z+1,i} = d_{z,i} - d_{z,y}$$

where

$$y = i - 2^z \pmod p \tag{3.9}$$

since each number of the form $np+i$ in the range 2^z to 2^{z+1} can be associated with a number of the $mp+y$ in the range 0 to 2^z. This is the basic recursion given by lines R.7 to R.10 of Fig. 3.2. The recursion can be started by direct computation of parity in the range 0 to 2^m, where $2^m < p < 2^{m+1}$ (lines R.3 to R.6 of the program). The calculation (in lines R.1 and R.2) of the exponent to which 2 belongs mod p, t, and of the powers of 2 mod p (they are the entries of the array r) obviates multiple-word calculation in the evaluation of (3.9) for large z (see line R.9).

EXERCISES

1. Let $R_i = \prod_{j=1 \text{ to } i}(m_j)$, $i = 1$ to n, where $m_1, m_2, \ldots, m_n$ are n given positive integers. The *mixed radix representation* of N, $0 \leqslant N < R_n$, is given by the digits $(d_0, d_1, \ldots, d_{n-1})$, where

$$N = d_0 + d_1 R_1 + d_2 R_2 + \ldots + d_{n-1} R_{n-1}$$

and

$$0 \leqslant d_i < m_{i+1}, \quad i = 0 \text{ to } n-1.$$

Write a program for computing these "mixed radix digits" for N (given in standard representation) in ascending order of subscripts.

2. That *every* $N < B^n$ may be expressed in the form (3.1) for $0 \le d_i < B$, $i \in n$, follows from the evident identity

$$\Sigma_{j=0 \text{ to } k-1}[(B-1)B^j] = B^k - 1.$$

What is the analogous identity for the expansion (3.2)?

3. List the Gray code defined by (3.4) for $n = 4$.

4. For relatively prime integers a and b, the last nonzero remainder produced by Euclid's algorithm (see exercise 6 of § 1.3) is 1. By successive substitution of expressions for the remainder r_i in terms of r_{i-1} and r_{i-2}, one may obtain a relation $ax + by = 1$ or $a(x+b) + b(y-a) = 1$. Use this approach to write an efficient *inverse(c,m)* procedure for use in evaluating (3.7).

*5. The residue representation of the integer 1 is $(1,1,\ldots,1)$. The residue representation for the multiplicative inverse of an integer b which is relatively prime to each modulus is thus

$$(inverse(b,m_1), inverse(b,m_2), \ldots, inverse(b,m_n)).$$

Using the approach of exercise 1, but performing the operations in residue representation, write a program to calculate the mixed radix digits, thence the standard representation for an integer N given in residue representation as $(a_1, a_2, \ldots, a_n)$. (Hint: $d_0 = a_1$. See Szabó and Tanaka [97] and Mann [88] for additional discussion and for variations on the conversion from residue to standard representation.)

6. (a) What is the residue representation of the number 1918 with respect to the primes 2, 3, 5, 7 and 11?

(b) The residue representation of the number $N < 2310 = 2 \times 3 \times 5 \times 7 \times 11$ is $(1,0,4,3,7)$. What is N?

7. Write a program to express $C = \binom{n}{r}$ in prime factor representation without actually factoring C.

(Hint: The exponent of a prime p in the prime factorization of $m!$ is

$$\Sigma_{i=1, 2, \ldots} \left(\left[\frac{m}{p^i} \right] \right).$$

(Also, see Comét [72].)

*8. (a) Why would the direct application of (3.8) be an inefficient method of calculating $d_{z,i}$?

(b) Evaluate $d_{z,0}$, $z = 0$ to 34, for the primes $p = 3, 7, 11$ and 17.

9. Using the identity

$$X^n \equiv X^{\Sigma_i \in set(n)^{(2^i)}} \equiv \prod_{i \in set(n)} \left(X^{(2^i)} \right)$$

write an efficient procedure, *power(X,n)*, which forms X^n for X real and n a whole number.

*10. The procedure suggested in exercise 9, though sufficient for practical purposes, is not the optimal method of forming X^n using only multiplications. For instance, X^{15} formed as $[(X^3)^2]^2 (X^3)$ requires only five multiplications, yet as $[(X^2)^2]^2 (X^4) (X^2) (X)$ it requires six. Let $R(n)$ be the minimum number of multiplications required to form X^n. Does $R(2n) = R(n) + 1$? (There are many interesting unanswered questions concerning $R(n)$ and its calculation. For further discussion of this and related problems, the reader is referred to Gioia *et al.* [80], Knuth [84], and to the "solution" to *American Mathematical Monthly* problem 5125 (August 1964).)

3.2. Sets and Vectors

Combinatorial computing involves the handling of various sets of things: sets of points, sets of permutations, sets of exponents. We distinguish between *unordered sets*, sets in which the order of presentation of the elements is incidental, and *ordered sets*,

sets in which the order of presentation of the elements is pertinent and explicitly stated. We generally use the unqualified word "set" to imply an unordered set (in fact, usually an unordered set of natural numbers) and the word "vector" to imply an ordered set.

3.2.1. *Unordered Sets—Masks*

The *bit-pattern* representation of unordered sets of natural numbers—i.e. quantities of our "set" data-type—is discussed in § 2.1.1. Unordered sets of objects other than natural numbers (or perhaps sparse sets of natural numbers for which the bit-pattern representation is uneconomical) call for a different representation.

One frequently useful approach is to work with the sets indirectly, to assign a distinct and easily calculable natural number index to each element of the universe of elements. Subsets of the universe are then represented by sets of indices and many operations on the original sets are reduced to basic set-theoretic operations on the sets of natural numbers. The key to effective use of this representation lies in the indexing scheme; since peculiar properties of the original elements often determine existence questions, it is important that conversion between elements and indices can be accomplished readily.

For the most part, indexing schemes are derived from formulae for enumerating elements of the given universe. For the elementary combinatorial objects—combinations, permutations, partitions, etc.—such formulae are known and provide a natural "serial number" indexing scheme. (Although not done in this text, it would be desirable for many algorithms to equate an object with its serial number automatically, letting the conversions be handled "behind the scenes" by a sophisticated software system.) Chapter 5 discusses serial number assignments for the common combinatorial entities.

By way of terminology, we often use the word *mask* to denote a set of natural number indices—a set whose elements specify a subset of a set of indexed objects. Masks are commonly used to indicate particular properties of certain objects of the indexed set, whether or not those objects have yet been "examined", for instance.

3.2.2. *Unordered Sets as Ordered Lists—Binary Search*

An alternative approach is simply to list the elements of a set as the entries of a 1-array. Set-theoretic manipulations of these "list-represented" sets are then programmed as procedures and referred to with mnemonic functional notation. (This representation should not be confused with the "*linked* list" representation used by many "list-processing" languages [51], even though such a representation indeed is sometimes useful. In fact, it is not difficult to envision a software system which would use assorted representations for a set (or, for that matter, an entity of any data-type), choosing as the calculation proceeds a representation consonant with the current properties of the set. The programmer, using only standard language, would then not be concerned with details of data-representation.)

The most commonly required operations on an unordered set are testing for the

existence of elements and adjoining new elements. When adjoin-type operations pre-
dominate, it is best to leave the list itself unordered, to append new elements at the end
of the list as they appear, and to test for element inclusion by direct search through the
list. However, when search-for-existence predominates (as it often does), then it is desirable to keep the list ordered so that a *binary search* (sometimes called a *logarithmic
search*) may be effected.

S.0   *adjoin*(s,"K","S")

S.1 $d = -1;\quad u = K$

S.2   until $d = u-1$:

S.3 $k = \left\lceil \dfrac{u+d}{2} \right\rceil$

S.4   if $s = S_k$: *adjoin*( ) = 1; exit from procedure

S.5   if $s < S_k$: $u = k$

S.6   otherwise: $d = k$

S.7   for $k = K-1, K-2, \ldots, u$: $S_{k+1} = S_k$

S.8 $S_u = s; K+1 \rightarrow K$

S.9   *adjoin*( ) = 0; exit from procedure

FIG. 3.3. A binary search procedure.

We present in Fig. 3.3 a procedure which searches for the element s in the ordered
list S_0 to $K-1$, $S_0 < S_1 < \ldots < S_{K-1}$, by the binary search technique. We first compare
the element s with an element near the middle of the list. As a result of this comparison
we then restrict the search either to the lower half or the upper half of the list. The next
comparison is made near the middle of the restricted range, which further "halves" the
range of search. This process continues until the element is found [*adjoin*( ) is set to 1] or
shown not to exist in the list [*adjoin*( ) is set to 0].

This procedure also includes program (lines S.7 and S.8) which adjoins s to the list
(with maintenance of the list ordering) when nonexistence has been shown. Of course,
these lines would be omitted if only information about existence were required. Other
modifications of this procedure are often desired; for instance, existence may call for
deletion of s from the set. Also, the comparisons of lines S.4 and S.5, which have been
indicated here as real number comparisons, may be involved comparison programs
operating on more complex elements of an unordered set. This situation often arises
with respect to the calculation of serial numbers for complex objects. With the objects
presorted (§ 6.2), the binary search technique is used to identify (i.e. calculate the serial
number for) a given arbitrary object of the set.

3.2.3. *Signatures—Fractional Word Representation*

An ordered set (a vector), because of the declared sequencing of its components, generally cannot be represented as compactly as an unordered set. It is quite natural to store the components of a vector of numbers in successive computer words—the vector is simply a 1-array of numbers. We do not have occasion to use the standard set-theoretic operations on these ordered sets: manipulations on vectors are programmed as entry-by-entry operations.

Following Lehmer [86], we employ the term *signature* to denote a vector of indices, a vector with integral components. It is sometimes convenient to store more than one component per computer word. We may express this fractional word model in set-theoretic language. For instance, if (a_0, a_1, a_2) is a signature with $0 \leqslant a_i < 58$, $i = 0$ to 2, then

$$A = set(a_0) \cup [set(a_1) \oplus \{6\}] \cup [set(a_2) \oplus \{12\}]$$

is a fractional word representation for (a_0, a_1, a_2). Only eighteen bits are needed for the bit-pattern representation of this set A.

Actually, representations of various sorts are simply signatures: the polynomial representation of a natural number is a signature of digits, the prime factor representation is a signature of exponents, and so forth. While the fractional word representation of such signatures is sometimes necessary for the sake of compactness, it often results in awkward manipulation of the numbers. Fractional word models of this sort are avoided. On the other hand, the use of a fractional word model can be expedient when "parallel" operations upon the components are required; examples of this appear in §§ 3.2.4 and 3.5.3.

Note that for a signature $V_{1 \text{ to } n}$,

$$\bar{V} = \bigcup_{i=1 \text{ to } n}(\{V_i\})$$

is the set of indices actually appearing in V. Since indices may be repeated in V, $|\bar{V}|$ can be less than n.

3.2.4. *Vectors and Polynomials Modulo 2*

Many algebraic studies call for handling of vectors and polynomials which have their components and coefficients, respectively, taken from a finite field, often from the field $\{0, 1, \ldots, p-1\}$ for p a prime. In most cases one merely stores each component or coefficient in a separate computer word since addition and multiplication require the isolation of operands. However, when $p=2$ so that the field is $\{0, 1\}$, addition and multiplication modulo 2 are symmetric difference and intersection of individual bits, hence a fractional word representation becomes economical.

With this representation, mod 2 vectors and polynomials are given as sets; for example, $v_{0 \text{ to } 5} = (0,0,1,1,0,1)$ is represented by $\{2,3,5\}$ and $p = x^6 + x^5 + x^3 + 1$ by $\{0,3,5,6\}$.

If u and v are two such vectors, their vectorial sum is simply $u\Delta v$; likewise, the polynomial sum of p and q is $p\Delta q$. The "scalar product" of u and v is $|u\cap v|$; the polynomial product of p and q is

$$\Delta_{i \in p}(q \oplus \{i\}). \tag{3.10}$$

(The notation "$p \otimes q$" would not be unnatural for this operation.)

As an example of mod 2 polynomial manipulation using this representation, we present in Fig. 3.4 a program for calculating all "irreducible" polynomials of degree $\leqslant 12$. A polynomial is *irreducible* if it is not the product of polynomials of lower degree. The algorithm is simply to test each polynomial Q of degree n for reducibility by seeing if there exists an (already computed) irreducible polynomial of degree $\leqslant n/2$ which divides Q. The test polynomial is formed at line I.4 (note that only polynomials with the terms x^n and x^0 are considered). The test itself is performed at line I.5. If all lower degree polynomials are prime to Q, Q is stored as the K_nth nth degree irreducible polynomial P_{n, K_n}, and the count of polynomials K_n is increased by one (line I.6). The *div*( ) procedure actually performs the polynomial division, returning 1 as a result if the division is exact and 0 otherwise. Mod 2 polynomial division is, of course, the inverse of multiplication given by (3.10). An analogue of the standard division algorithm (see exercise 11 of § 2.2) is used here. Note (line D.3) that for these polynomials addition is its own inverse—subtraction is again the symmetric difference operation.

The reader is referred to Elspas [74] and Peterson [92] for discussions of applications and related topics.

I.1   for $n = 1$ to 12:

I.2 $K_n = 0$

I.3   for $i = 0, 1, \ldots, 2^{n-1}-1$:

I.4 $Q = (set(i) \oplus \{1\}) \cup \{0,n\}$

I.5   if $\forall_{n*=1 \text{ to } n/2; \, k=0 \text{ to } K_{n*}-1}(div(Q,P_{n*,k}) = 0)$:

I.6 $P_{n,K_n} = Q; \quad K_n+1 \rightarrow K_n$

I.7   "At this point the irreducible polynomials have been computed."

I.D.0 $div(f,g)$ "Does f/g have 0 remainder?"

I.D.1 $a = (f)^1; \, b = (g)^1$

I.D.2   until $a < b$:

I.D.3 $f \, \Delta[g \oplus \{a-b\}] \rightarrow f$

I.D.4   if $f = \{ \ \} : div(\ ) = 1$; exit from procedure

I.D.5 $a = (f)^1$

I.D.6 $div(\ ) = 0$

FIG. 3.4. Calculation of irreducible modulo 2 polynomials.

EXERCISES

1. Write procedures for performing the union and intersection operations on unordered sets given as ordered lists.

2. For compactness the set of odd primes $\{3,5,7,11,\ldots,p\}$ is often represented by the mask $P = \{0,1,2,4,\ldots,(p-3)/2\}$ where an element $j \in P$ yields a prime by the conversion $2j+3$. What is the conversion associated with the still more compact representation of the set $\{5,7,11,13,\ldots,p\}$ in which no space in the bit-pattern is reserved for the (odd) multiples of 3?

*3. Calculate all integral solutions to the Diophantine equation

$$\binom{x+r-1}{r} + \binom{y+r-1}{r} = \binom{z+r-1}{r}$$

for $1 \leqslant x \leqslant y \leqslant z \leqslant 10{,}000$ and $r = 3, 4, 5$. (Hint: Use the binary search procedure on precalculated tables—see Wunderlich [102].)

4. The binary search procedure is an optimal method for locating an *arbitrary* entry in an ordered list of numbers, i.e. on the average it will find an entry with as few array references as possible. Given a "convex unimodal" list $S_{1 \text{ to } N}$, $S_1 < S_2 < \ldots < S_m$, $S_m > S_{m+1} > \ldots > S_N$, the corresponding optimal procedure for locating the unique maximum point m is based on the property of the Fibonacci sequence u_n—see (1.5)—illustrated in the following diagram:

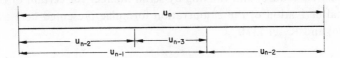

Write a procedure for locating this maximum, with as few references as possible, from the arbitrary convex unimodal list $S_{1 \text{ to } N}$. (For proofs and further reading on this subject see Bellman and Dreyfus [5].)

5. A mask M, whose elements fall in isolated blocks, is sometimes conveniently given as a union of intervals by the 1-array $(v_0,v_1,v_2,\ldots,v_{2m+1})$, where $M = \bigcup_{i=0,2,\ldots,2m} (v_{i+1} \sim v_i)$ and $v_j < v_{j+1}$, $j = 0$ to $2m$, for instance. Write procedures for accomplishing the union and intersection of masks represented in this way.

6. Let the two vectors $(u_1,u_2,\ldots,u_n)$ and $(v_1,v_2,\ldots,v_n)$ with components from $\{0,1,2\}$ be represented by the sets $U1$, $U2$ and $V1$, $V2$ respectively where $U1 = \bigcup_{i=1 \text{ to } n \ni u_i=1}(\{i\})$, $U2 = \bigcup_{i=1 \text{ to } n \ni u_i=2}(\{i\})$, etc. Write an expression for the scalar product modulo 3 of these vectors—i.e. for $\sum_{i=1 \text{ to } n}(u_iv_i) \pmod 3$—in terms of their representation as sets.

7. Consider polynomials with coefficients taken from the finite field $\{0,1,\ldots,p-1\}$, p an odd prime.

   (a) Assuming fractional word representation of the vector of coefficients (use m bits for each component where $2^{m-1} < p < 2^m$) write a program to add two such polynomials.

   (b) With the coefficients given as standard 1-arrays of numbers, write a program to multiply two polynomials.

   (c) Using the fact that $a^{p-1} = 1 \pmod p$—Fermat's theorem [34]—hence a^{-1} is $a^{p-2} \pmod p$ in our field, write a program to divide polynomial A by polynomial B. That is, write a program to form polynomials Q and R, where $A = BQ+R$, degree $R <$ degree B.

8. Let $Ser(x)$ be the serial number for x in X and $Ser(y)$ be that for y in Y. Devise a serial number assignment for elements in the set $X \times Y$ (the set of all ordered pairs with elements of X for the first component and elements of Y for the second component).

9. Write a program which associates a unique number $Ser(p,q)$ with a given positive rational number p/q. Assume $gcd(p,q) = 1$. (Hint: Recall the proof for the countability of the rational numbers.)

10. (a) Write a function *index(i,a,n)*, using the binary search technique, which produces

$$\text{first } k \in n \ni a_k = i$$

as its result. Assume that $a_0 < a_1 < \ldots < a_{n-1}$ and that

$$|\bigcup_{k \,\in\, n \,\ni\, a_k = i}(\{k\})| = 1$$

is always true.

(b) Describe a situation in which it would be necessary to form

$$index[index(x,A2,N2), A1, N1].$$

11. Write a set of procedures to add, subtract, multiply, and divide kth degree polynomials in one variable. Note that for multiplication and division the result will in general be truncated. For division assume that the divisor has a constant term.

3.3. Elementary Combinatorial Configurations[†]

The study of combinations and permutations usually serves as one's introduction to the realm of combinatorial mathematics. These and other elementary mathematical objects such as partitions and compositions are fundamental to combinatorial computing. Definitions and basic representations are presented in this section; Chapter 5 considers generation procedures and ranking by serial number for certain of these configurations. Formal derivation of the enumeration formulae presented here may be found in Riordan [36] and Ryser [37].

3.3.1. *Combinations and Permutations*

A *combination* of n (distinct) objects taken r at a time, an *r-combination of n elements*, is a selection of r of the n objects without regard to order. When the selection is made without replacement (i.e. when repetitions of the objects are not allowed), the r-combination is a subset of the n objects which contains exactly r elements—it is an *r-subset* of an *n-set*. When, further, the objects are natural numbers, an r-combination is simply an unordered set containing r natural numbers. A compact representation for such a combination is the bit-pattern representation of the corresponding subset, i.e. as $\{c_0,c_1,\ldots,c_{r-1}\}$. For example, recalling from § 2.1.1 that bit-patterns are basically read from right to left, we see that the bit-pattern 0101100 represents the combination $\{2,3,5\}$ for $n = 7$.

There are $\binom{n}{r}$ r-subsets of a given n-set and 2^n subsets altogether.

Alternate interpretations of a given bit-pattern give rise to related representations. For instance, by reading the above pattern backwards one has the combination $\{1,3,4\}$, and by interchanging the roles of 0-bits with 1-bits, one obtains the 4-combination $\{0,1,4,6\}$. These representations are conveniently called the *reverse* and *dual* representations respectively. They arise as natural transformations in many problems.

[†] The word "configuration" is generally used as a synonym for "entity" and not in any technical sense [21], but see § 7.3.

When repetition of objects is allowed, the available representations are less compact. The basic bit-pattern form represents the r-combination

$$(c_0, c_1, \ldots, c_{r-1}), \quad 0 \leqslant c_{i-1} \leqslant c_i \leqslant n-1 \text{ for } i \in r, \tag{3.11}$$

by the set

$$\bigcup_{i \in r}(\{c_i + i\}). \tag{3.12}$$

For example, (1,2,2,6) is given by the bit-pattern 1000011010. This representation requires $n+r-1$ bits. From the one-to-one correspondence between such bit-patterns containing r 1-bits and r-combinations with object-repetition allowed, one sees that there are $\binom{n+r-1}{r}$ such r-combinations of n objects.

Of course the basic signature form (3.11)—with the restriction $c_{i-1} < c_i$ when object-repetition is not allowed—is available when bit-manipulation is not convenient.

A *permutation* of n objects taken r at a time (an *r-permutation of n elements*) is an *ordered* selection of r of the n objects. When explicit reference to r is not made, we assume that $r = n$. The n objects being permuted may or may not be distinct. In either case, a permutation is just a vector of objects and may be represented as such. In the common case where the objects are distinct natural numbers, a permutation is a signature with distinct components, say $(p_0, p_1, \ldots, p_{n-1})$ with $p_i \neq p_j$ when $i \neq j$. For compactness it is sometimes necessary to apply the fractional word representation to permutations. This, however, is generally wasteful of time as a permutation in this form is more difficult to analyze (but see exercise 1).

There are $n(n-1)(n-2) \ldots (n-r+1)$ r-permutations of n distinct objects—when $r = n$, this reduces to $n!$. The number of permutations of n not necessarily distinct objects depends on the "specification" of the objects. When the objects are classified into k, $k \leqslant n$, groups, each group containing m_i, $i = 1$ to k, equal objects, then there are

$$\frac{n!}{m_1! \, m_2! \, \ldots \, m_k!} \quad \left(\text{where } \sum_i (m_i) = n\right) \tag{3.13}$$

distinguishable permutations of the n objects. Note that if $k = n$ so that $m_i = 1$ for $i = 1$ to k, then this also reduces to $n!$.

3.3.2. *Permutation as Transformations*

In combinatorial computing, permutations of natural numbers are used both in the static sense defined above (i.e. as an existing rearrangement of numbers) and in the operational sense as a rule describing a transformation upon some class of objects (i.e. as an element of a group). For the latter interpretation, the elements of the object-set are labeled with distinct natural numbers (usually 0 to $n-1$, or 1 to n), and the permutation is a vector with these indices as components, the components themselves subscripted by the same set. Thus the permutation $(p_0, p_1, \ldots, p_{n-1})$ indicates that, for $i = 0, 1, \ldots, n-1$, the object in position i is replaced by the object in position p_i; alternatively, the object in position p_i moves to position i. For example, the result of applying the permutation $(2,1,3,0)$ to four objects stored as the vector (b_0,b_1,b_2,b_3) is the vector $(b_2,b_1,b_3,b_0) \equiv (\bar{b}_0,\bar{b}_1,\bar{b}_2,\bar{b}_3)$.

If permutation $(p_0,p_1,\ldots,p_{n-1})$ is applied and then permutation $(q_0,q_1,\ldots,q_{n-1})$ is applied to the new vector (i.e. $q \cdot p$—see next section), the result is the same as if

$$(p_{q_0},p_{q_1},\ldots,p_{q_{n-1}}) \tag{3.14}$$

were applied to the original arrangement of objects. The permutation inverse to $(p_0,p_1,\ldots,p_{n-1})$ is $(p_0^*,p_1^*,\ldots,p_{n-1}^*)$, where $p_{p_i}^* = p_{p_i^*} = i$ for $i = 0$ to $n-1$. (With each permutation $p_{0 \text{ to } n-1}$ one may associate a *permutation matrix*, a matrix of zeros and ones in which row i contains a single one in column p_i. Under this association permutation composition corresponds to matrix multiplication and permutation inversion to matrix transposition.)

In discussing the effect of a permutation as a transformation it is convenient to express the permutation as a product of "cycles". A *cycle* of a permutation $(p_0,p_1,\ldots,p_{n-1})$ is a vector $(p_{a_1},p_{a_2},\ldots,p_{a_r})$ where $a_i = p_{a_{i-1}}$, $i = 2$ to r, and $a_1 = p_{a_r}$. It is customary to omit the commas between components of this vector when labels not requiring separation are used. Thus, for example, $(0\ 2)$ is a cycle of $(2,3,0,1)$. Every permutation can be expressed uniquely (up to order of the cycles and initial component of each cycle) as a product of disjoint cycles [9]. For example, $(2,1,5,4,3,0) \equiv (0\ 2\ 5)(1)(3\ 4)$.

Apart from its use as an expository tool (see § 5.2.2) and in group theory [9] and combinatorial analysis [36], this representation of a permutation as a product of cycles can be useful in programs which require transformation of several object-sets according to the same permutation (see § 6.2.3). A procedure for calculating disjoint cycles $c_{0,\ -}, c_{1,\ -}, \ldots$ of a permutation $p_{0 \text{ to } n-1}$ is given in Fig. 3.5. When the calculation is complete, N gives the number of cycles and j_k, $k = 0$ to $N-1$, gives the number of components of the cycle labeled k.

```
C.0    cycle(p,"c")
C.1        (real array) p
C.2        S = (set) n
C.3        for i ∈ S, as N = 0, 1, …:
C.4            c_{N,0} = p_i
C.5            for k = p_i, …, p_k → k, …, until k = i, as j_N = 1, 2, …:
C.6                S ~ {k} → S
C.7                c_{N,j_N} = p_k
           "calculation is complete"
```

FIG. 3.5. Calculation of cycle structure for a permutation.

3.3.3. *Finite Groups*

As in combinatorics, the primary role of the group concept in combinatorial computing is to unify the treatment of symmetry, hence to assist in the enumeration of "inequivalent" (i.e. representative) configurations (§ 6.6). Also, interesting combinatorial problems arise from the intrinsic study of groups as algebraic systems (see Forsythe [76] and Trotter [98], for instance).

We recall that an *abstract* (finite) *group* may be defined as a finite set of elements *closed* under an *associative multiplication* operation which contains a *unique* (multiplicative) *identity* element and in which each element possesses a *unique* (multiplicative) *inverse*. We are mainly concerned with groups of transformations, in fact with groups of permutations where multiplication is the composition operation defined by (3.14): if p and q are permutations of an object-set, then $q \cdot p$ is the permutation obtained by first applying p to the object-set and then q to that result. One can easily see that this multiplication is associative, hence to show that a set of permutations is a group, one need only establish the closure, identity, and inverse properties. We do not require very deep group theoretic results in this book; further concepts and characteristics are presented as needed.

A group of transformations is most often given by a set of "generators", a set of transformations whose iterated products and powers constitute the group. For example, (now using the labels 1,2,3,4) the permutation (2,3,4,1) is a generator for the "cyclic" group $\{(1,2,3,4),(2,3,4,1),(3,4,1,2),(4,1,2,3)\}$, while the permutations (3,4,1,2) and (2,1,4,3) generate the group

$$\{(1,2,3,4), \quad (3,4,1,2), \quad (2,1,4,3), \quad (4,3,2,1)\}. \tag{3.15}$$

A set of generators or in fact a group of permutations itself is an unordered set of permutations and is represented as such. (Because of the closure property, the permutations of a group generate themselves.) The problem of constructing a group from a given set of generators is pursued in § 6.5.3.

An abstract group of *order* n (i.e. the group has n elements) may be given by an $n \times n$ "multiplication table", an array $G_{1 \text{ to } n, 1 \text{ to } n}$ in which the entry $G_{i,j}$ specifies the product of the elements indexed by i and j. A group multiplication table has very special properties which are used in certain combinatorial investigations of groups. Label the elements of the group $1, 2, \ldots, n$, letting 1 represent the identity element; denote the product of i and j by ij (not necessarily equal to ji) and the inverse of i by i^{-1}. With the row index denoting the left factor and the column index the right factor of a product, we may summarize the properties of a group multiplication table as follows:

(1) Each row and each column is a permutation of $(1,2,\ldots,n)$.
(2) Row 1 and column 1 are the identity permutation $(1,2,\ldots,n)$.
(3) The patterns

$$
\begin{array}{cccc}
i \cdots ij & 1 \cdots i & j \cdots 1 & ji \cdots j \\
\vdots \quad \vdots & \vdots \quad \vdots & \vdots \quad \vdots & \vdots \quad \vdots \\
1 \cdots j & j \cdots ji & ij \cdots i & i \cdots 1
\end{array}
$$

are valid for all "squares" which contain the identity at one of its corners.
(4) If the group is *commutative* (i.e. if $ij = ji$ for all i and j in the group), then the table is symmetric with respect to the NW–SE diagonal.

These properties follow from the group axioms. For example, the first pattern of (3) is derived as follows: Suppose the edges of the square are in rows a and b and in columns c and d so that $ac = i$, $bc = 1$, and $bd = j$. Then $ad = a \cdot 1 \cdot d = a(cc^{-1})d = (ac)(c^{-1})d = (ac)(bd) = ij$.

An abstract group is sometimes given by a set of generators $g_1, g_2, \ldots, g_m$ and a set of relations of the form $g_1^{e_1} \cdot g_2^{e_2} \cdot \ldots \cdot g_m^{e_m} = 1$ (the e_h, $h = 1$ to m, are arbitrary integers) which the generators must satisfy. For example, the generators a and b which satisfy the relations $a^2 = 1$ and $b^2 = 1$ define the group $\{1,a,b,ab\}$ which is isomorphic to the group given by (3.15).

Actually an arbitrary set of relations need not define a proper finite group. The relations may be inconsistent and thereby yield equivalence relations between the generators themselves or they may be incomplete and define an infinite group. This interesting problem of deducing group structure from a given set of relations is discussed by Leech [85] and by Trotter [98].

A representation convenient in work of this kind is to use the symbols 0, 2, 4, ... for the group elements $g_0, g_1, g_2, \ldots$ (including generators, of course) and 1, 3, 5, ... for their corresponding inverses. A relation is then merely an ordered set of positive integers, terminated, perhaps, with the identity symbol. For instance, the signature (3,6,4,4,0) represents the relation $q_1^{-1}g_3g_2^2 = 1$. (Note that both 0 and 1 represent the identity element of the group.)

3.3.4. *Compositions and Partitions of a Whole Number*

A *composition* of a whole number N is a signature $(C_1, C_2, \ldots, C_m)$ which satisfies the conditions

$$C_i > 0, \quad i = 1 \text{ to } m,$$
and
$$C_1 + C_2 + \ldots + C_m = N. \tag{3.16}$$

The components of the signature are said to be the *parts* of the composition. A composition with m parts is an *m-part composition*. Note that the order of the parts is significant; the composition (1,6,2,2) is distinct from the composition (6,1,2,2).

A *partition* of N is a composition in which the order of the parts is irrelevant. Thus a partition may be expressed as a signature provided a convention regarding the order of the components is established. Doing this we say a partition of N is a signature $(P_1, P_2, \ldots, P_m)$ which satisfies the conditions

$$P_i > 0, \quad i = 1 \text{ to } m,$$
$$P_1 + P_2 + \ldots + P_m = N, \tag{3.17}$$
and
$$P_1 \geqslant P_2 \geqslant \ldots \geqslant P_m.$$

We sometimes allow parts of a partition (or composition) to be zero so that the first restriction of (3.17) is replaced by $P_i \geqslant 0$ (or $C_i \geqslant 0$). However, when no qualification is made, we are assuming *non-zero* parts. As far as their representation is concerned, compositions are related to partitions as permutations with repeated objects are related to combinations with repeated objects—the relevance or irrelevance, respectively, of the order of the parts makes the difference.

The representation of compositions and partitions as signatures satisfying (3.16) and (3.17) respectively is certainly the most common. In certain cases, however, there do exist more compact representations, not less amenable to analysis than these *standard representations*.

A one-to-one correspondence between compositions of N and subsets of $\{0,1,\ldots,$ $N-2\}$ may be established as follows: With $S \subset \{0,1,\ldots,N-2\}$ where $|S| = m-1$, associate the composition $(C_1,C_2,\ldots,C_m)$ where $C_i = (S)_i - (S)_{i-1}$ for $i = 1$ to m [use $(S)_0 = -1$ and $(S)_m = N-1$]. With $(C_1,C_2,\ldots,C_m)$ associate the set

$$\bigcup_{i=1 \text{ to } m-1} \left(\left\{ \sum_{j=1 \text{ to } i}(C_j) - 1 \right\} \right).$$

Using the bit-pattern representation for sets, this affords a compact *difference represen-tation* for compositions. For example, the composition (7,2,2,1,4) of 16 is represented by the pattern 000110101000000. The value of N, which cannot be discerned from the bit-pattern, must be stored separately.

This subset-composition correspondence shows that there are $\binom{N-1}{m-1}$ m-part compo-

sitions of N and 2^{N-1} compositions in all. A short table of the number of partitions of N with size of parts $\leqslant M$ is given in Appendix I; its derivation is discussed in § 5.5.2.

The difference representation can also be used for partitions; for example, the pattern 1100100010000 represents the partition (5,4,3,1,1). We must remember, however, that because of the inequalities of (3.17) many patterns (e.g. 011001001) do not represent partitions.

Another bit-pattern model for partitions is the so-called *rim representation* suggested by Comét [73]. With a partition $(P_1,P_2,\ldots,P_m)$ of N we associate the set S defined by

$$S = (N+1) \sim \bigcup_{i=1 \text{ to } m} (\{P_i + m - i\}). \tag{3.18}$$

For $j < k$, we have $P_j \geqslant P_k$, hence $P_j + m - j > P_k + m - k$. Therefore S contains $N+1-m$ elements. The m integers not in S correspond to the m parts, the size of a part being given by the number of elements in S smaller than the corresponding integer. Specifically,

$$P_j = |((N+1) \sim S)_{m+1-j} \cap S|$$

or, alternatively,

$$((N+1) \sim S)_j = P_{m+1-j} + j - 1.$$

In bit-pattern language, the number of 1-bits to the right of the jth 0-bit (counted from the right) is the value of P_{m+1-j}. For example, the representation of (5,2,2,1,1,1) is 1101110010001. In practice, since N is known, the leftmost 1-bits of this representation are extraneous.

The rim representation is easily visualized in relation to the *Ferrers graph* of a partition. Figure 3.6 gives a Ferrers graph for the partition (5,2,2,1,1,1) and shows that in traversing the bottom and right side *rim* of the diagram, a 1-bit is inserted in the representation for each left-to-right increment of the path and a 0-bit for each bottom-to-top increment.

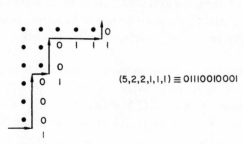

FIG. 3.6. Ferrers graph and rim representation.

A *multiplicity representation* for partitions, which resembles the prime factor representation for integers, is sometimes useful. Here a partition is given as a set of parts and corresponding signature of multiplicities. For example, the partition (5,2,2,1,1,1) is given as the set $\{1,2,5\}$ along with the vector (3,2,1). The jth component of the vector indicates the number of times the jth element of the set appears as part of the partition. By allowing zero multiplicities, the set can be understood as $\{1,2,\ldots,N\}$ and need not be explicitly expressed.

The reader is referred to Macmahon [29], Section IV, for additional discussion of mathematically useful representations of compositions and partitions.

3.3.5. *Partitions of a Set*

A *partition of a set S* is a family of disjoint subsets of S whose union is S. A generally satisfactory representation, which arises from the straightforward generation of the partitions of a set of natural numbers (see § 5.5), consists of presenting a partition as a 1-array of subsets ordered according to the smallest element of each subset. That is, a partition is given by an array of sets $P_{1 \text{ to } r}$, where

$$(P_1)_1 < (P_2)_1 < \ldots < (P_r)_1. \tag{3.19}$$

Here r is the number of subsets (i.e. number of parts) of the partition. An alternate representation suggested by Hutchinson [82], useful when bit-manipulation is not available, is as a vector $p_{1 \text{ to } n}$, where $p_i = j$ when $i \in P_j$ as given by (3.19). Here we have $S = \{1,2,\ldots,n\}$.

The number of partitions of a set containing n elements is B_n, where

$$B_n = \sum_{k=0 \text{ to } n-1} \left[\binom{n-1}{k} B_k \right], \quad B_0 = 1. \tag{3.20}$$

A small table of these *exponential* (or *Bell*) numbers is given in Appendix I; the truth of (3.20) is most easily seen following the discussion of the generation scheme in § 5.5.

EXERCISES

1. Let $(p_0, p_1, \ldots, p_{n-1})$ and $(q_0, q_1, \ldots, q_{n-1})$ be two permutations of the natural numbers 0, 1, $\ldots, n-1$. Devise a representation of these permutations as sets so that the number of "coincidences" of the permutations may be calculated efficiently (a *coincidence* is a value of i, $i \in n$, such that $p_i = q_i$). (Hint: Consider linearizations of permutation matrices.)

2. Let $p_{0 \text{ to } n-1}$ be a permutation of the numbers 0 to $n-1$. Write a program to calculate the permutation $p^*_{0 \text{ to } n-1}$ which is inverse to p.

3. A permutation is even or odd according as the number of its cycles (in its representation as a product of disjoint cycles) with an even number of components is even or odd, respectively [9]. For instance, the identity permutation which contains no even cycles is even and a transposition which contains precisely one even cycle is odd. Write a program to calculate the parity of a given permutation.

4. Let $c_{0,1 \text{ to } j_0}$, $c_{1,1 \text{ to } j_1}$, $\ldots$, $c_{N-1,1 \text{ to } j_{N-1}}$, be disjoint cycles of a permutation $(p_0, p_1, \ldots, p_{n-1})$. Write a program to calculate $p_{0 \text{ to } n-1}$.

5. Let $C_{0 \text{ to } c-1}$, $\_$ and $D_{0 \text{ to } d-1}$, $\_$ be the cycle representations for the permutations p and q of (0,1, $\ldots, n-1$), respectively. Write a program to calculate $B_{0 \text{ to } b-1}$, $\_$, the cycle representation for the product $q \cdot p$ (i.e. p is applied and then q—be careful not to get the inverse).

6. Discuss the relationship between the cycle representation for permutations of $(0,1, \ldots, n-1)$ and partitions of the set $\{0,1, \ldots, n-1\}$.

7. How many m-part compositions of N are there when void parts are allowed?

8. What is the dual, reverse 4-combination of $\{0,1,2,3,4,5\}$ associated with $(1,1,1,1,3,4)$ which is a 6-combination of $\{0,1,2,3,4\}$? Object-repetition is allowed, of course.

9. Write programs to convert between the standard signature representation and rim representation of a partition.

10. Let $M_{1 \text{ to } n, 1 \text{ to } n}$ be the multiplication table for a group. For group elements i, j, and k, write formulae for the following elements: $i \cdot j \cdot k$, i^{-1}, $j \cdot (i \cdot k)^{-1}$.

11. What relationship do the elements i and j possess if the square

$$
\begin{array}{ccccc}
1 & \cdot & \cdot & \cdot & i \\
\cdot & & & & \cdot \\
\cdot & & & & \cdot \\
\cdot & & & & \cdot \\
j & \cdot & \cdot & \cdot & 1
\end{array}
$$

exists in a multiplication table for a group?

12. A *semi-group* is a finite set of elements closed under an associative multiplication. What properties (1) to (4) of the text remain valid for the multiplication table of a semi-group?

13. The rim representation applies as well to combinations as to partitions. List the bit-patterns for the twenty 3-combinations of the objects 0, 1, 2, 3 with repetitions allowed. (Since each pattern contains exactly three zeros, ones to the left of the third zero need not be written.)

14. The *conjugate partition* of a given partition is the partition associated with the transposed Ferrers diagram of the given partition; for instance, (6,3,1,1,1) is conjugate to (5,2,2,1,1,1). Write programs (or expressions) which form the conjugate partition for a partition given in the standard, difference, rim, and multiplicity representations. (See also exercise 19 of § 5.5.)

15. The *Frobenius representation* [87] gives a partition as a pair of vectors each with r components where r is the number of dots along the main diagonal of the Ferrers graph. The components of the first vector give the number of dots in the rows above the main diagonal while those of the second vector give the number of dots in the columns below the diagonal. For example, (6,2,1) and (4,2,0) represent the partition (7,4,4,2,1). Can these vectors be themselves represented as unordered sets? Write programs to convert from the standard representation to an efficient version of the Frobenius representation, and vice versa.

16. List the fifteen partitions of the set $\{0,1,2,3\}$ in each of the representations discussed in the text

3.4. Linear Graphs and Networks

A connectivity relationship (i.e. a binary relation) between elements of an arbitrary set of objects may be conveniently and directly expressed by presentation of the set of pairs of objects which possess the relationship. A "linear graph" is the mathematical abstraction of such an expression. Linear graphs have been extensively studied by pure mathematicians interested in their inherent combinatorial properties and by applied mathematicians concerned with them as models for electrical networks, maps, and other real situations. Graphs and associated "zero-one (incidence) matrices" are fundamental to combinatorial computing as the basic means of expressing relations among elements of a finite set.

(The subject of graph enumeration is not treated in this text. The reader is referred to Riordan [36] for an introduction to this important aspect of combinatorial analysis.)

3.4.1. *Definitions and Terminology*

A *linear graph* or just *graph* is an abstract mathematical configuration consisting of a set of elements called *vertices* or *nodes* and a set of pairs of these elements called *edges* or *links*. We are concerned in this book with *finite graphs*, graphs in which both the set of vertices and the set of edges is finite. The n vertices of a graph are usually labeled with the numbers $0, 1, \ldots, n-1$. An edge of the graph is then represented by an ordered or unordered pair of these numbers (i, j) or $\{i, j\}$, i, and $j \in n$. (See exercise 1.)

An edge is *directed* or *undirected* according as the pair of vertices is ordered or unordered. A graph containing only directed (respectively undirected) edges is a *directed* (respectively *undirected*) graph. A graph may be *partly directed*. (It is often convenient to view an undirected edge as being directed in both directions. In this context, a partly directed graph is simply a directed graph.) The first vertex of a directed edge is said to *precede* the second vertex, the second *succeeds* the first. The two vertices of an undirected edge are *adjacent* to each other; a successor is *adjacent* to its predecessor. Pictorially, vertices are customarily represented by points, undirected edges by lines connecting the points and the orientation of a directed edge by an arrowhead pointing from the predecessor to successor vertex—see Fig. 3.7.

Both undirected and directed graphs may have *loops* (not to be confused with "iterations"), edges of the form $\{i\}$ or (i,i) which connect a vertex to itself. Also, vertices may exist which are not part of any edge of the graph; such vertices are called *isolated vertices*. Of course, by definition, the vertices of an edge of a graph must be vertices of the graph. The *valence* or *degree* of a vertex is the number of edges which contain that vertex as a component—a loop contributes 2 to the valence. The valence of a vertex v in a directed graph consists of the sum of its *in-valence*, the number of edges containing v as a second component, and its *out-valence*, the number of edges containing v as a first component.

By defining a graph as a *set* of edges we do not admit "multiple (i.e. parallel) edges",

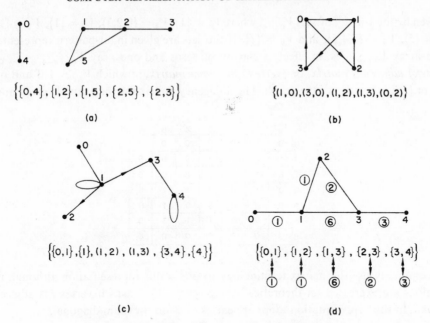

FIG. 3.7. Sample (labeled) graphs and networks.

i.e. identical ordered pairs of vertices considered to be distinct edges of the graph. However, many applications prescribe that a real number weight factor be associated with each edge; such a weighted graph is often called a *network* (more precisely a "capacitated" network). Actually, it is perhaps best to consider a network as having only directed edges and, in symmetric problems, to adopt the viewpoint discussed above.

A *subgraph* of a given graph G is a graph whose vertices and edges are respectively subsets of the vertices and edges of G. Additional graph-theoretic definitions are given as needed and in the exercises at the end of this section.

Figure 3.7 gives examples of four graphs and networks along with pictorial representations. Figure 3.7(a) is simply an undirected graph; Fig. 3.7(b) is a directed graph; Fig. 3.7(c) is a partly directed graph containing loops; and Fig. 3.7(d) is a symmetric network, the weights being circled and written close to the corresponding edges.

3.4.2. *Representation by Incidence Matrices*

The most generally useful model for a linear graph is the *vertex adjacency representation* in which the set of adjacent (or successor) vertices is given for each vertex of the graph. For example, the graph

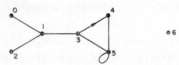

is given by the 1-array of sets $V_{0 \text{ to } 6}$, where $V_0 = \{1\}$, $V_1 = \{0,2,3\}$, $V_2 = \{1\}$, $V_3 = \{1,4,5\}$, $V_4 = \{5\}$, $V_5 = \{3,4,5\}$, and $V_6 = \{\ \}$. If the sets are given in bit-pattern representation, the 1-array $V_{0 \text{ to } n-1}$ is, in effect, a 2-array of zeros and ones, say $V_{0 \text{ to } n-1, 0 \text{ to } n-1}$—the *(vertex) adjacency matrix*, or *(vertex) incidence matrix*, in which $V_{i,j} = 1$ if and only if (i,j) or $\{i,j\}$ is an edge of the graph. The adjacency matrix for the example given above is

	0	1	2	3	4	5	6
0	0	1	0	0	0	0	0
1	1	0	1	1	0	0	0
2	0	1	0	0	0	0	0
3	0	1	0	0	1	1	0
4	0	0	0	0	0	1	0
5	0	0	0	1	1	1	0
6	0	0	0	0	0	0	0

We commonly use the matrix terminology to imply this representation although manipulations are expressed set-theoretically, e.g. "if $j \in V_i$:" asks if vertex j is adjacent to vertex i. In this representation loops appear as ones on the main diagonal.

Note that the adjacency matrix for an undirected graph is symmetric, i.e. $j \in V_i \Leftrightarrow i \in V_j$. In general, very little is gained in attempting to take advantage of this symmetry by deleting "half" of the matrix—we always assume a complete square array. For directed graphs it is sometimes convenient to work with the "transpose" of V (the 1-array $V^*_{0 \text{ to } n-1}$, where $j \in V_i \Leftrightarrow i \in V^*_j$), or, due to the technical difficulty of transposing a bit matrix (see § 6.5.1), to have available both the adjacency matrix and its transpose.

For a network, the incidence matrix is a 2-array of real numbers, the entry in the ith row and jth column giving the weight for the link (i,j). There are two common ways in which the nonexistence of links in a network may be indicated. One is with the weights themselves; for instance, a weight of zero (or inversely an infinite—i.e. extremely large—weight) can denote the absence of that link from the network. Alternatively, a vertex adjacency matrix can be associated with the network matrix of weights. This second alternative is perhaps more compatible with the set-theoretic features of our language—see § 4.3.2.

A computer model useful in many applications is the *edge–vertex incidence representation* in which a graph containing e edges is given by a 1-array of sets $E_{0 \text{ to } e-1}$, where E_k, $k \in e$, contains as elements the indices for the two vertices of the edge with index k. The entries of E can be thought of as the edges of the graph. If these entries are given in bit-pattern representation, then E is essentially an $e \times v$ 2-array of zeros and ones, *the edge–vertex incidence matrix*. Note that each row of this matrix has precisely two nonzero elements unless that row corresponds to a loop, in which case the row contains a single such element.

This representation cannot be used *per se* for directed or partly directed graphs since we are using unordered sets to represent the edges. However, such a representation may be expanded in several ways to allow for directed graphs. One way is to associate with E a mask M whose elements are the indices of the edges for which the natural order of the vertex indices gives the proper direction of the edge, i.e. M is the union of singleton sets $\{k\}$ for which $(E_k)^1$ succeeds $(E_k)_1$. For networks, an additional array $W_{0 \text{ to } e-1}$ containing the weights must be available.

Another representation immediately applicable only to undirected graphs is the *edge incidence representation*. Here a graph G is given by an e by e symmetric matrix M of zeros and ones—$M_{h,k} = 1$ if and only if edge h and edge k have a common vertex. Although of theoretical interest, this matrix seems to be of little practical value.

3.4.3. *Incidence Systems*

The abstract concept of a linear graph has a direct "generalization" of considerable importance in combinatorial computing. An *incidence system* consists of a finite set of elements called *objects* or *varieties* and a finite collection of subsets of these elements called *blocks*. Incidence systems in which objects are distributed among the blocks with a definite regularity are called block designs—they are extremely important in combinatorial mathematics. (A particular block design is discussed in § 6.4.2.) In general, however, no restrictions are placed on the blocks: a block may be empty, two distinct blocks may contain precisely the same objects, any object may occur in an arbitrary number of blocks, and so forth. Manipulations with general incidence systems are frequently required in combinatorial computing; the more common ones are discussed in §§ 6.3 and 6.4.

It is convenient to label the objects with the numbers $0, 1, \ldots, n-1$ and the blocks with numbers $0, 1, \ldots, m-1$. An incidence system may then be represented by a 1-array of sets $B_{0 \text{ to } m-1}$, where $B_i \subset \{0,1,\ldots,n-1\}$, i.e. object j is in block i if and only if $j \in B_i$. With the bit-pattern representation for the sets, this again yields a zero–one incidence matrix, the *block–object incidence matrix*. It is sometimes useful to abuse the terminology slightly by equating the objects with their labels, i.e. to consider the numbers $0, 1, \ldots, n-1$ as the objects themselves. Then the representation becomes the incidence system itself. As with graphs, it is often useful to have available an incidence matrix and its transpose, the 1-array of sets $V_{0 \text{ to } n-1}$, where $i \in V_j$ if and only if $j \in B_i$.

Treating the edges as blocks and the vertices as objects, a graph can be considered an incidence system. Conversely, treating both the blocks and objects as vertices, but allowing edges only between blocks and objects, an incidence system can be considered as a *bipartite graph* [28].

3.4.4. *Tree-like Data Structures*

A *path* in a directed graph is a sequence of edges $(i_0,i_1), (i_1,i_2), (i_2,i_3), \ldots, (i_{k-1},i_k)$ in which no edge appears more than once. Similarly, a sequence of distinct edges $\{i_0,i_1\}$, $\ldots, \{i_{k-1},i_k\}$ is a path in an undirected graph. A path is a *cycle* if $i_0 = i_k$. (Note that these definitions allow repeated vertices. For precision of terminology, the words "arc" and "circuit" are sometimes used to imply no repeated vertices, although there is certainly no unanimity in the literature in this respect.) The vertices contained in the edges of the path are "connected" by the path. An undirected graph is *connected* if every pair of its vertices are connected by some path. A *tree* (sometimes called "free tree") is a connected undirected graph containing no cycles. An *oriented tree* is a tree in which each edge has been assigned a unique direction. (By contrast, a "directed tree" allows cycles with two edges.)

An *arborescence* is an oriented tree in which a single node, the *origin*, has zero in-valence, while all other nodes have in-valence equal to one. Those nodes with zero out-valence are called *terminal nodes*. Pictorially we represent an arborescence as a "root system", with the origin at ground level (level zero) and the other nodes arranged below at levels one, two, three, etc., according to their distance from the origin. We omit the arrowheads since all edges are pointing downward. Any node of an arborescence is the origin for the sub-arborescence, called the *branch* or *limb*, emanating from the node, consisting of all nodes below and connected to the given node.

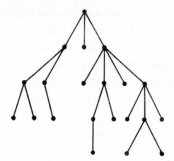

FIG. 3.8. Sample arborescence.

A simple four-level arborescence is pictured in Fig. 3.8. (Besides the usual directed graph terminology—predecessor, successor, etc.—it is convenient to use family-tree terminology when discussing an arborescence. Words like "father", "son", "brother", "ancestor" appear in the text without formal definition.)

The concept of an arborescence emerges in combinatorial computing primarily with respect to the structure of combinatorial algorithms; this subject is pursued in Chapter 4. Occasionally, however, the data themselves are or need to be structured in this fashion. Consider the arborescent representation of the mathematical expression

$$a[(ac)/[d-log(f)]+g]-proc(b,a,c) \tag{3.21}$$

given in Fig. 3.9. Each terminal node corresponds to a variable and each nonterminal node to an operation whose operands are its successor nodes. A nonterminal node also represents the result of its operation which, except for the origin, is an operand for its predecessor. There are, of course, many possible "linear" representations of such an expression, (3.21) being one. A partially analyzed canonical form used frequently in formula translation programs is the so-called "Polish notation" in which the tree is

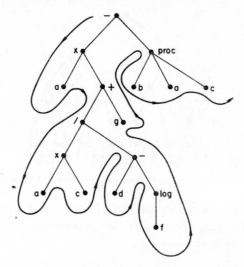

FIG. 3.9. Arborescent representation of mathematical expression.

reduced to a list of symbols, void of grouping symbols, by a left-to-right scan as illustrated by the arrows of Fig. 3.9. The list

$$-, \times, a, +, /, \times, a, c, -, d, \log, f, g, \textit{proc}, b, a, c$$

is a Polish form of the expression (3.21), for instance. Of course, the number of operands associated with each operator must be known. Furthermore, due to the associativity and commutativity of some of the operators, additional ordering conventions are required to establish the Polish notation as a unique representation.

(A significant area of high-speed data processing has grown up with respect to the representation and processing of tree-like structures which contain symbolic information at their nodes. This so-called "list-processing" touches, and even overlaps, combinatorial computing as presented in this book at several places. However, the language and implementation of list-processing, as it stands today, is more oriented towards symbol manipulation and is less suitable for the expression and efficient solution of mathematical problems. We refrain from delving into this interesting and closely related area of computing in this book; the reader is referred to Fox [78] and Knuth [27] for an introduction to this branch of nonnumerical computing.)

In combinatorial computing it is sometimes necessary to represent only the framework of a reasonably uniform arborescence. Indeed, the structure of an arborescence with the out-valence of every node $\leqslant 2$—a binary tree—can be represented compactly. Consider the "full" binary tree and node labeling of Fig. 3.10. Every such arborescence with depth $\leqslant 4$ is a *subtree* of that binary tree, hence is given by a subset of $\{1, 2, \ldots, 31\}$. Furthermore, this node labeling permits easy location of a particular node given its label. For instance, node 22 ($= 10110$ in binary) is at level 3 of the tree since $4 - (set(22))_1 = 3$; it may be reached from the origin by taking successively a right branch since

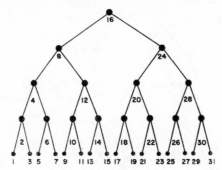

Fig. 3.10. A full binary tree with common node labeling.

$set(22) \cap \{4\} \neq \{ \}$, a left branch since $set(22) \cap \{3\} = \{ \}$, and then a right branch since $set(22) \cap \{2\} \neq \{ \}$. The generalization of this representation to arbitrary "regular" arborescences is pursued in the exercises.

Exercises

1. What vertex label permutations leave the given representation for the graph of Fig. 3.7(a) unchanged? (The problem of detecting distinct representations of a graph is pursued in § 7.1. Furthermore, in enumeration problems one must usually be careful to distinguish between a *labeled graph* and a *graph*.)

2. The complement, say $\bar{G}$, of a graph G has the same vertex set as G, while its edge set contains all those pairs of vertices which are not edges of G.

(a) Write a program to compute $\bar{G}_{0 \text{ to } v-1}$ when $G_{0 \text{ to } v-1}$ is the vertex adjacency model for G.

(b) Write a program to compute $\bar{G}_{0 \text{ to } \bar{e}-1}$ when $G_{0 \text{ to } e-1}$ is the edge–vertex incidence representation.

$$\left(\text{Note that } e + \bar{e} = \binom{v}{2}. \right)$$

3. A subgraph H is a *section subgraph* of a given graph G if every pair of vertices contained in H which is an edge of G is also an edge of H. Let $G \, (= G_{0 \text{ to } n-1})$ be given in adjacency representation. Write a program to determine if an array of sets H_i for $i \in I$, $I \subset n$, represents a section subgraph of G.

4. The *star* of a subset of vertices V of a graph G is a subgraph containing all edges which have an element of V as a component (and all vertices of those edges, of course). Write a program to form the smallest section subgraph which contains the star of a given set of vertices—assume G is given as an edge–vertex incidence matrix.

5. The *complete graph* with n vertices contains all possible pairs of distinct vertices. A k-clique is a complete subgraph with k vertices; these are $\binom{k}{2}$ edges in an undirected clique and $2\binom{k}{2}$ in a directed one. Write a program which determines whether an arbitrary subset of vertices K, $K \subset n$, is a K-clique of a given undirected graph $G_{0 \text{ to } n-1}$.

6. Construct the block–object incidence matrix associated with the incidence system formed with the 3-subsets of $\{0,1,2,3\}$ as the blocks. Labeling the rows and columns 0, 1, 2, 3, draw the graph which results from viewing the above matrix as an adjacency matrix.

7. Given a block–object incidence matrix $B_{0 \text{ to } m-1}$, $B_i \subset n$, one may define the *block–block* incidence matrix $B^*_{0 \text{ to } m-1}$, $B^*_i \subset m$, by $j \in B^*_i \leftrightarrow B_j \cap B_i \neq \{ \}$. Write a program to accomplish this construction. (A block–block matrix can, of course, be an adjacency matrix for a graph. It is known (see Erdős [75]) that the inverse construction is also possible, i.e. given an $m \times m$ adjacency matrix, one can find an n, $n \leq [m^2/4]$, and a set of blocks B_i, $i \in m$, with $B_i \subset n$, such that the block–block incidence matrix equals the given matrix.)

8. A *matroid* (see [83]) is an incidence system $B_{0 \text{ to } m-1}$ satisfying the properties (1) no block is a proper subset of another, and (2) if $x, y \in n$ and $x \in B_i \cap B_j$ and $y \in B_i \sim B_j$, then there exists a $k \in m$ such that $x \notin B_k$, $y \in B_k$, and $B_k \subset B_i \cup B_j$ (equality permitted). Write a program which tests whether or not a given incidence system is a matroid.

9. A *full regular tree* has n link-levels ($n+1$ node-levels counting the level of the origin) and exactly m nodes emanating from each nonterminal node. Devise a node indexing scheme for a full regular tree which would permit a compact set-theoretic representation of an ordered arborescence which is a sub-tree of the regular tree.

10. Devise an indexing scheme for full binary trees in which nodes at the same level are assigned sets of consecutive indices. (Hint: This can be done so that the node indices have a simple set-theoretic relationship with the node indices assigned by the scheme given in the text.)

3.5. The n-Cube

Many practical problems involve, in their mathematical formulation, manipulations with the so-called *n-cube*. This important topological configuration consists of a collection of 0-*cells* (vertices), 1-*cells* (pairs of vertices—edges), 2-*cells* (quadruples of vertices), and so forth, all related in a definite manner. Specifically, let $V =$ (set) 2^n be the set of vertices. An edge consists of a pair of these natural numbers $\{u,v\}$ in which the binary expansion of u differs in a single-bit position from the binary expansion of v (e.g. the pair $\{9,13\} = \{1001,1101\}$ is an edge of the 4-cube). Altogether there are $n \times 2^{n-1}$ edges in the n-cube, since there is a choice of n distinct bit-positions in which two vertices may be made to differ and for each of these choices there are 2^{n-1} ways in which the remaining $n-1$ bits may be assigned values 0 or 1. In general, a d-cell consists of 2^d vertices, those vertices with a particular $n-d$ bits fixed in value: there are $\binom{n}{d} 2^{n-d}$ d-cells and $3^n = \sum_{d=0 \text{ to } n} \left[\binom{n}{d} 2^{n-d} \right]$ cells of all dimensions in the n-cube. A cell of the n-cube is quite naturally represented by a number containing n ternary (i.e. base 3) digits, the twos corresponding to bits *not* fixed in value. For example, the signature of digits $(0,1,2,2,0)$ represents the 2-cell $\{01000,01010,01100,01110\}$. An interesting fractional precision representation for such a signature, where the digits used are 1, 2, and 3 in place of 0, 1, and 2, is employed in § 3.5.3.

Although the n-cube is considered to be the collection of all of its cells, it is common to present the n-cube merely as the set $\{0,1,\ldots,2^n-1\}$, the connections between the vertices being understood. By a *subset of the n-cube* we mean a subset of $\{0,1,\ldots,2^n-1\}$ along with those cells all of whose vertices belong to the subset. A subset of the n-cube can thus be represented compactly as a bit-pattern. Although we say that certain d-cells, $d > 0$, are in the subset, they are not explicitly represented and must be computed (see Fig. 3.11).

3.5.1. *Graph-theoretic Representation*

Ignoring all cells except vertices and edges, the n-cube is simply a *regular* graph of degree n (i. e. a linear graph with precisely n edges incident with each vertex). As many studies are concerned only with graph-theoretic properties, it is useful to be able to construct graph-theoretic models of the n-cube. These constructions are not difficult; for instance, the iteration

$$\text{for } i = 0 \text{ to } 2^n - 1: \quad G_i = \bigcup_{j \, \in \, n} \left(\{ nbr(set(i) \triangle \{j\}) \} \right) \tag{3.22}$$

produces the adjacency matrix G for the n-cube. The symmetric difference operation is used here to complement the jth bit of the bit-pattern $set(i)$.

3.5.2. *Normal Form Expressions of Boolean Functions*

A variable is said to be *Boolean* or *logical* if it may assume only one of two possible constant values, which we may take to be 1 and 0 (often representing "true" and "false" respectively). The vector $(x_0, x_1, \ldots, x_{n-1})$, where $x_0, x_1, \ldots, x_{n-1}$ are n independent Boolean variables, has 2^n possible values. Each of these values may be associated uniquely with a vertex of the n-cube in an obvious manner (for instance, the vector $(x_0, x_1, \ldots, x_{n-1})$ with the vertex $v = x_0 \cdot 2^{n-1} + x_1 \cdot 2^{n-2} + \ldots + x_{n-1}$). Therefore, a *Boolean function* (a function whose *range* has only two values 0 and 1) of n Boolean variables is a mapping of the n-cube into the 2-cube, and may be represented by the subset of vertices of the n-cube which map into the value 1. There are 2^{2^n} distinct Boolean functions of n variables since that is the number of subsets of vertices of the n-cube.

A common and useful means of presenting a Boolean function is by means of a *sum-product Boolean expression*, i.e. a mathematical expression involving Boolean variables and constants which uses only the unary operation of complementation (denoted by priming and defined by $0' = 1$ and $1' = 0$) and the two binary operations of logical addition (denoted with a plus symbol and defined by $0+0 = 0$ and $0+1 = 1+0 = 1+1 = 1$) and logical multiplication (denoted with a centered dot or by juxtaposition and defined by $0 \cdot 0 = 0 \cdot 1 = 1 \cdot 0 = 0$ and $1 \cdot 1 = 1$). There are many different sum-product expressions for a single Boolean function, e.g. $x_0(x_1 + x_2')$ and $x_0 x_1 x_2 + x_0 x_2'$ are two expressions for the function of three variables represented by $\{4,6,7\}$.

Consider now the 2^n functions of n variables corresponding to the expressions

$$x_0^{i_0} x_1^{i_1} x_2^{i_2} \ldots x_{n-1}^{i_{n-1}}, \tag{3.23}$$

where $i_j \in \{0,1\}$ for $j = 0$ to $n-1$, and $x_k^0 \equiv x_k'$ and $x_k^1 \equiv x_k$. Since this expression has the value 1 only when $x_j = i_j$, $j = 0$ to $n-1$, each of these functions is represented by a single vertex of the n-cube, the vertex whose binary representation is $i_0 i_1 \ldots i_{n-1}$. As any function of n variables is the union of singleton sets of vertices, any function may be expressed as the logical sum of terms of the form (3.23). Such an expression is called the *developed normal form* of the function.

Two expressions like (3.23) which correspond to adjacent vertices differ only in a single superscript; for example, vertex 2 ($= 010$) and vertex 6 ($= 110$), which correspond to expressions $x_0' x_1 x_2'$ and $x_0 x_1 x_2'$ respectively, are adjacent vertices in the 3-cube. We may say that a function which contains adjacent vertices contains the edge defined by those vertices, and, in fact, we may associate a particular expression with this edge.

In the above example, if vertices 2 and 6 are both in the function, so is the edge $\{2,6\}$; the associated expression is $x_1 x_2' (= x_0 x_1 x_2' + x_0' x_1 x_2')$. In general, from repeated application of the identity

$$x_0^{i_0} x_1^{i_1} \ldots x_j^{i_j} \ldots x_{n-1}^{i_{n-1}} + x_0^{i_0} x_1^{i_1} \ldots \left(x_j^{i_j}\right)' \ldots x_{n-1}^{i_{n-1}} \equiv x_0^{i_0} x_1^{i_1} \ldots x_{j-1}^{i_{j-1}} x_{j+1}^{i_{j+1}} \ldots x_{n-1}^{i_{n-1}}$$

we see that a d-cell of the n-cube corresponds to a product of $n-d$ distinct primed or unprimed variables—a *fundamental formula*. We say that a d-cell is contained in a function if all of its vertices are in the function (the corresponding fundamental formula *implies* the function).

A sum–product expression of a Boolean function which is a sum of fundamental formulae is called a *normal form* expression for the function. While a given Boolean function has a unique developed normal form, it has many normal form expressions. Selecting a "simple" normal form expression for a given Boolean function is a significant combinatorial problem pursued in §§ 6.4 and 7.5 as well as in the following section.

3.5.3. *Prime Implicants (Basic Cells)*

Let f be a Boolean function of n variables, or, equivalently, a subset of the n-cube. A cell contained in f is a prime implicant (sometimes called a basic cell—see Prather [35]) of f if it is not a proper subset of any other cell contained in f. It is easily seen that the collection of prime implicants for a given function is unique. The importance of the prime implicants of a function is due to the fact, first proved by Quine [94], that every "minimal" normal form expression consists of the logical sum of fundamental formulae corresponding to prime implicants (see § 7.5).

Given a function of n variables as a set of vertices, the prime implicants may be formed by the iterative process of building successively larger and larger cells which are still contained in the function, the process terminating when the largest cells—the prime implicants—have all been formed. A program incorporating this process is given in Fig. 3.11. For representing a d-cell (hence a prime implicant) we use a fractional word representation of the signature of ternary digits (using digits 1, 2, and 3). For example, the 1-cell $\{10110, 10100\}$ which can be given by the signature $(1,0,1,2,0)$, or equivalently by $(2,1,2,3,1)$, is represented by the bit-pattern (i.e. set) 1001101101 where each component of the signature occupies two bits of the pattern. This model is convenient since moderately sized cells generally occupy a single computer word. Also, the union of two "opposite" $(d-1)$-cells is directly a d-cell (this property is used at line P.8 of the program).

Cells of dimension d are stored in rows $C_{d,\_}$ of the 2-array C. The number of such cells is N_d. Line P.1 (and the *cell*( ) procedure) models the 0-cells (the vertices) from the function f, given as a subset of vertices, and from n, the number of variables. The body of the loop whose quantifier is on line P.2 forms all $(d+1)$ cells from existing d-cells. Those d-cells which are used to form a $(d+1)$-cell are recorded as not being prime implicants (line P.9); thus p_d is a mask which ultimately gives those d-cells which *are* prime

"Given n and f"

P.1	for $v \in f$, as $N_0 = 0, 1, \ldots$: $C_{0,N_0} = cell(\ )$
P.2	for $d = 0, 1, \ldots$, until $N_d = 0$:
P.3	$N_{d+1} = 0$; $p_d = $ (set) N_d
P.4	for $h = 0$ to N_d-2; $i = h+1$ to N_d-1:
P.5	$C^* = C_{d,h} \Delta C_{d,i}$
P.6	$x = (C^*)_1$
P.7	unless x is even and $C^* = \{x, x+1\}$: reiterate
P.8	$C^* = C_{d,h} \cup C_{d,i}$
P.9	$p_d \sim \{h,i\} \rightarrow p_d$
P.10	if $d = 0$ or $\bigvee_{j=0 \text{ to } N_{d+1}-1} (C^* \neq C_{d+1,j})$:
P.11	$C_{d+1,N_{d+1}} = C^*$; $N_{d+1}+1 \rightarrow N_{d+1}$

"At this point the prime implicants are given by"

"$C_{0,(p_0)_1}, C_{0,(p_0)_2}, \ldots; C_{1,(p_1)_1}, C_{1,(p_1)_2}, \ldots; \ldots;$"

"$C_{d-1,(p_{d-1})_1}, C_{d-1,(p_{d-1})_2}, \ldots$"

P.C.0	$cell(\ )$	
P.C.1	$C = \{\ \}$	
P.C.2	for $i = 0$ to $n-1$, as $V = set(v), \ldots, V \ominus \{1\} \rightarrow V, \ldots$:
P.C.3	$(C \oplus \{2\}) \cup set(nbr(V \cap \{0\})+1) \rightarrow C$
P.C.4	$cell(\ ) = C$; exit from procedure

FIG. 3.11. Calculation of prime implicants.

implicants. The quantifier at line P.4 controls the search for $(d+1)$-cells. Two d-cells are *opposite faces* of a $(d+1)$-cell if their symmetric difference (line P.5) "spans" a single dimension (if the test of line P.7 is unsuccessful). These d-cells are then not prime implicants (line P.9); they form a $(d+1)$-cell (line P.11). Except for edges, $(d+1)$-cells may be formed from several pairs of d-cells—the test of line P.10 insures that duplicate cells are not listed.

EXERCISES

1. Write a program to compute all d-cells (as sets of vertices) of the n-cube.

2. Let f be a subset of the vertices of the n-cube. Write a program to compute the vertex adjacency matrix for f.

*3. A collection of d-cells $c_0, c_1, \ldots, c_{m-1}$ is *irredundant* if $\bigcup_{i \in m} (c_i) \neq \bigcup_{i \in m \sim \{j\}} (c_i)$ for each $j \in m$. Given c_i, $i = 0$ to $m-1$, in the peculiar representation used for cells in Fig. 3.11, write a program to compute an irredundant sub-collection of cells.

4. The set of prime implicants of a function f yields a unique representation of f as a sum of products of "literals" (primed or unprimed variables). How may one express f uniquely as a product of sums of literals?

*5. Verify by experiment that the number of prime implicants of the "average" Boolean function of n variables exceeds 2^{n-1} first for $n = 9$. (Formulae for this number are given in Wells [101] and Mileto and Putzolu [90].)

6. Write a program to form all subsets s_i, $i = 1, 2, \ldots$, of $\{0,1,\ldots,n-1\}$ which have the following two properties: (1) if $x \in s_i$ and $y \in s_i$, then $x-y \not\equiv 1 \pmod{n}$, and (2) each s_i is "maximal", that is, no j exists, $j \neq i$, such that $s_i \subset s_j$.

3.6. Geometric Configurations

Programming a computer to play board-games such as chess and to solve various geometric puzzles involves a realization of "physical" shapes and situations within the computer. We present here a few remarks on the representation of certain two-dimensional rectilinear configurations.

3.6.1. *Polyominoes*

Several amusing combinatorial problems (see § 7.4) are concerned with "polyominoes" (the terminology is due to Golomb [18])—connected plane figures composed of equal-sized squares joined along their edges. We speak of dominoes, trominoes, tetrominoes, $\ldots$, n-onimoes according to the number of "cells" (squares) in the figure. A polyomino is unchanged by rigid motions in space; the polyominoes for $1 \leqslant n \leqslant 4$ are given in Fig. 3.12.

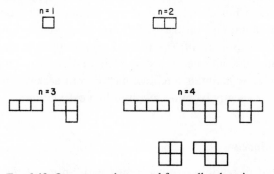

FIG. 3.12. One-, two-, three-, and four-cell polyominoes.

A polyomino in a fixed position in space can be given by the coordinates of its cells relative to an established coordinate system. Because of the economical bit-pattern representation for sets, these coordinates are best expressed as a 1-array of sets of x-coordinates; the heptomino in Fig. 3.13(a) could be represented by $P_0 = \{1,2\}$, $P_1 = \{2\}$, $P_2 = \{2\}$, $P_3 = \{1,2,3\}$, for example. Of course, this representation is not unique: the array of sets

$$P_0 = \{0\}, \quad P_1 = \{0,1,2,3\}, \quad P_2 = \{0,3\} \tag{3.24}$$

also models this heptomino. To establish a unique representation one must place the

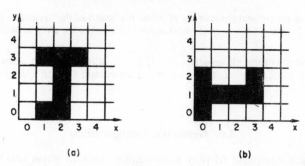

FIG. 3.13. A heptomino and its canonical representation.

polyomino in a canonical position in space. This may be done by choosing among all bit-pattern representations the one for which $P_0 \neq \{\ \}$, $\exists_i(0 \in P_i)$, and P is minimum (recall from § 1.4.5 the manner in which arrays of sets are compared). The canonical representation of the heptomino is shown in Fig. 3.13(b).

When working within a restricted area, on an "$n \times m$ chessboard", for instance, it can be advantageous to consolidate the entries of the array into a single set. For example, our heptomino could be represented as

$$\{0\} \cup (\{0,1,2,3\} \oplus \{12\}) \cup (\{0,3\} \oplus \{24\}), \quad 12 \geqslant n, \tag{3.25}$$

instead of as in (3.24).

The primary manipulations on polyominoes (or similar plane figures) are motions in space: translations and horizontal, vertical and diagonal reflections—see § 6.4. Translations and the horizontal and vertical reflections are accomplished conveniently, using either representation, with the shift ($\oplus$ and $\ominus$) operations. The diagonal reflection (i.e. transposition) is more difficult, hence it is sometimes desirable to have the polyominoes available in two forms, to maintain a representative from each of the two "transposition classes" of figures for each polyomino involved in a calculation.

3.6.2. Chessboard Models

Many games and recreations are concerned with an $n \times m$ chessboard—a rectangle subdivided into n rows (ranks) and m columns (files) of small squares by lines drawn parallel to its sides. Such a configuration may be given by a 2-array $B_{1 \text{ to } n, 1 \text{ to } m}$ of course, but the nature of many inquiries calls for modifications to this basic representation.

When only occupancy of the squares is involved, it is often advantageous to present the board as a 1-array of masks, $b_{0 \text{ to } n-1}$, entry b_i indicating by its elements the columns in row i which are occupied (or unoccupied). A related variation is concerned with indicating the extent of the chessboards. In playing chess, e.g. $n = m = 8$, a knight's move consists, say, of adding 2 to the file index and subtracting 1 from the rank index. Rather than be burdened with comparison of the new index values with the limits 8 and -1 each time such a move is made, it is sometimes better to maintain an "unoccupiable border" around the board. For example, the array $b_{0 \text{ to } 11}, b_0, b_1, b_{10}, b_{11} = \{ \ \}$, $b_{2 \text{ to } 9} = \{2,3,4,5,6,7,8,9\}$, indicates the extent of a chessboard (with shift of index range, of course) by existence of elements in its sets.

It is occasionally desirable to view the chessboard other than as a *rectangular* array of squares. The diagonal moves of the bishop and queen in a game of chess, for instance, call for more elaborate representations in an attempt to endow the computer with the "sight" of the board possessed by a human. Having indexed the individual squares of the board, a list of diagonals given as occupancy-masks, cross-indexed with the squares they contain, is one possible representation in this respect. For intricate calculations (chess playing in particular) the chessboard may be given by many special representations besides the basic rectilinear form.

We close this chapter with an example showing the construction of the graph-theoretic model of the $N \times M$ chessboard where squares of the board are vertices of the graph and edges of the graph connect squares a knight's move apart. (A knight moves between opposite "ends" of the polyomino $P_0 = \{0\}$, $P_1 = \{0\}$, $P_2 = \{0,1\}$, which may be placed at any point on the board.) In Fig. 3.14, $G_{0 \text{ to } 63}$ is the resulting graph given as an adjacency matrix.

$$s = 0$$

for $n = 0$ to $N-1$:

    for $m = 0$ to $M-1$, as $s = s, s+1, \ldots$:

        $G_s = 0$

        for $Dn = -2, -1, \ldots, 2$ except when $Dn = 0$:

            $n^* = n + Dn$; if $n^* < 0$ or $n^* \geqslant N$: reiterate

            for $Dm = -2, -1, \ldots, 2$ such that $|Dn| + |Dm| = 3$:

                $m^* = m + Dm$; if $m^* < 0$ or $m^* \geqslant M$: reiterate

                $G_s \cup \{n^* M + m^*\} \rightarrow G_s$

FIG. 3.14. Construction of "knight-tour" graphical representation of chessboard.

With regard to this and similar constructions, however, it should be kept in mind that it is sometimes desirable to work directly from the actual physical situation rather than to form an abstract mathematical model. For example, the knight's tour problem (moving a knight about an $n \times n$ chessboard without landing on the same square twice (see § 4.2.4), which can be expressed as a graph-theoretic problem, is perhaps more efficiently programmed directly. Such cases arise when the physical situation has special properties—e.g. extreme symmetry—which might be "lost" (i.e. difficult to extract) when a general model is used. On the other hand, one should be aware of the existence of a general representation in order that established manipulations of the model can be translated, when needed, to the non-abstracted terminology of the original problem.

EXERCISES

1. Draw pictures for the twelve pentominoes in the canonical position described in the text.

2. "Interval subsets" (i.e. sets of the form $j \sim k$ for natural numbers j and $k, j > k$) may be extracted from a given set S by a process similar to Lehmer's method for counting 1-bits (§ 2.2.3); for example,

$$S \sim set[nbr(S) + 2^{(s)_1}]$$

is the "smallest" interval subset of S. Incorporate this process into a program which determines whether an arbitrary 1-array of sets represents a polyomino. (Squares touching at a corner only are not considered connected, although they may be part of the same polyomino due to a chain of edgewise connections elsewhere.)

3. Rewrite the program of Fig. 3.14 using the bordered model approach for delimitation of the extent of the board.

4. Some games and puzzles are concerned with only "half" of a chessboard, with the squares whose coordinate sum is odd basally connected along the diagonals. Write a program to convert such an "$n \times n$ checkerboard" into an $(n-1) \times n$ chessboard with squares missing from each corner; give the chessboard as a 1-array of sets $b_{0 \text{ to } n-2}$ with elements of the set indicating existing cells of the board.

*5. A (nondegenerate) Gaussian prime is a complex number $a + b\sqrt{-1}$, a and b positive integers, where $a^2 + b^2$ is an ordinary prime.

   (a) Write a program to form the array of sets $G_{1 \text{ to } m}$ where $x \in G_y$ if and only if $x + y\sqrt{-1}$ is a Gaussian prime. (Assume the set of ordinary primes P is given.)

   (b) Interpreting the checkerboard on which the primes (except for $1 + \sqrt{-i}$) lay as a rectilinear chessboard (see exercise 4), count the n-ominoes, $n = 1, 2, 3, \ldots$, on the board which do not abut the "right" or "upper" edge of the board (i.e. ignore n-ominoes containing the primes $x + y\sqrt{-1}$, where $x = m$ or $y = m$). (Hint: The approach of exercise 2 can be used to find the polyominoes.)

6. (a) Two-dimensional tick-tack-toe is played on a two-dimensional board $B_{1 \text{ to } 3, 1 \text{ to } 3}$. Labeling the squares of this board $0, 1, \ldots, 8$ in a natural manner, list the eight sets of three elements which serve as winning "lines" on the board. Label these sets $0, 1, \ldots, 7$ and form the sets L_s, $s = 0$ to 8, which indicate the lines in which square s is contained.

   (b) Write a program to construct a similar representation for three-dimensional tick-tack-toe.

*7. Let P_0 be a polyomino near the center of a large chessboard B. Define recursively P_i, $i = 1, 2, \ldots$, by the rules $P_i = P_{i-1} \cup Q$, where Q consists of all squares in $B \sim P_{i-1}$ which have exactly one edge in common with P_{i-1}. Write a program to study the growth pattern for each of the five tetrominoes. (Ulam [99] and others have studied the effect of *various* such simple rules for the growth of polyomino-like figures.)

CHAPTER 4

SEARCH AND ENUMERATION—BACKTRACK PROGRAMMING

Most combinatorial algorithms involve the generative enumeration of one or all configurations satisfying certain requirements. If the requirements are stringent, the enumeration becomes a search, the purpose of which might be to establish the existence or non-existence of a particular configuration. A program for realizing a systematic search or enumeration frequently, and quite naturally, takes the form of a nest of iterations. A search logically so structured has been called "backtrack" by Lehmer [121] and "backtrack programming" by Golomb and Baumert [111]. (Although we use established terminology in this book, we would slightly prefer the more descriptive and perhaps more general term "tree programming".) Such systematic and exhaustive procedures were employed on electronic computers as early as 1953, although the first computerized formalization of the "method" was by Walker [133] in 1958.

Because of its brute-force character and the difficulty of performing desk analysis on resulting algorithms, backtrack programming is often unfairly considered crude and used only as a last resort. Actually, in practice, a backtrack algorithm incorporating well-designed "impasse detection" (§ 4.2) is frequently the most direct and efficient approach to an enumerative calculation. This chapter uses the concept of nested iteration and its associated with-statement notation as a basis for a detailed discussion of backtrack programming—a fundamental tool of combinatorial computing.

4.1. Introduction to Backtrack Programming—The Search Tree

Consider the classification of the nine 2-digit base 3 numbers d_1d_0 by means of the arborescence shown in Fig. 4.1. Each link emanating from the origin corresponds to one of the three possible values for the digit d_1. Having chosen a value for that digit, there are then three choices for the digit d_0 indicated by the three links emanating from each first-level node. This simple example illustrates the "search tree" corresponding to the nest

$$\text{for } d_1 = 0 \text{ to } 2:$$
$$\text{for } d_0 = 0 \text{ to } 2:$$
$$\text{PROGRAM}$$

93

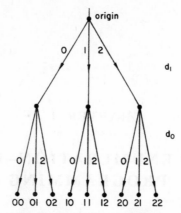

FIG. 4.1. Search tree for base 3 number generation.

which generates the nine numbers $d_1 d_0$. It is often helpful to visualize the structure of search and enumeration programs by these search trees—arborescences (§ 3.4.4) in which the left-to-right order of the branches is relevant.

In order to maintain a close terminological association between nests of iterations and search trees, we visualize our trees with downward growth, using words such as "shallow" and "deep" ("depth") in place of the more customary "low" and "high" ("height"). The downward-pointing arrows are often omitted in the figures.

4.1.1. *Elementary Signature Generation*

The generation of vectors with integral components that satisfy simple conditions provides an excellent introduction to backtrack programming and the search-tree visualization. Consider first the generation of the factorial digits a_i, $i = 0$ to $n-1$, where $0 \leqslant a_i \leqslant i$ (see § 3.1.1). This may be accomplished with the program of Fig. 4.2 whose associated search tree for $n = 4$ is given in Fig. 4.3. Note that each value of the nest index corresponds to one level of the tree and that the links emanating from a given node correspond to possible values for the iteration index at a certain point in the generation.

This example illustrates a search tree for a particular, and quite uniform, nest of iterations. In general, a search tree may have a nonuniform structure: the number of links emanating from nodes at the same level may be different, or the depth (i.e. number

$$\text{with } i = n-1, n-2, \ldots, 0:$$

$$\vdots$$

$$\text{for } a_i = 0 \text{ to } i:$$

$$\vdots$$

$$\text{USE } (a_0, a_1, \ldots, a_{n-1})$$

FIG. 4.2. Factorial digit generation.

of levels) of various branches may not be equal. Although we are always concerned with finite search trees, the exact structure of a tree may be unknown to the researcher. In fact, it is often the goal of a search program to gather information about the search tree — for example, to count the number of terminal nodes. Further, while we usually try to prearrange a calculation so that the total size of the search tree is minimized, many times it takes information gathered *during* the search to let us prune branches from the tree, hence to shape the exact course of the calculation. This is characteristic of many search programs and of combinatorial computing in general.

A level of nodes in a search tree corresponds to a particular iteration level within the nest, hence to a particular value of the nest index. Although we speak of level zero, level one, ..., level N of the tree, the example of Fig. 4.2 shows that we do not imply that the nest index always has value $0, 1, \ldots, N$ at those levels. A nest quantifier may assign values to the nest index in a general way (see § 1.5.2). Indeed, the manner in which the nest index is assigned values strongly influences the structure of the search tree. On the other hand, different nests of iterations may correspond to the same search tree. For instance, the tree for the lexicographic permutation generation given in § 5.2.1 (for $n = 4$) has a structure identical to that displayed in Fig. 4.3.

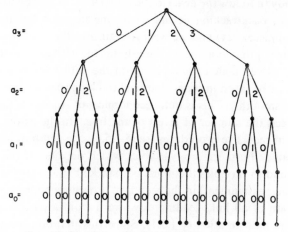

FIG. 4.3. Search tree for factorial digit generation.

Signature generation of a less uniform nature is illustrated in Fig. 4.4. This program generates vectors $(C_1, C_2, \ldots, C_m)$ subject to the restrictions that $C_1 + C_2 + \ldots + C_m = n$ and $0 < C_i \leqslant n$ for $i = 1$ to m; that is, it generates m-part compositions of n. Again note that the depth of the nest and its associated tree is controlled by the nest quantifier while the "breadth", the number of links emanating from the various nodes, is controlled by the general iteration quantifier. This particular generation is discussed more fully in § 5.4.2—indeed, much of Chapter 5 is concerned with signature generation.

with $i = 1$ to $m-1$, as $s = 0, 0+C_{i-1}, \ldots$:

$$\text{for } C_i = 1 \text{ to } n+i-m-s:$$

$$C_m = n-s$$
$$\text{USE } (C_1, C_2, \ldots, C_m)$$

FIG. 4.4. Program for m-part composition generation.

4.1.2. *Normal Tree Scan and Basic Program Structure*

Relative to the search-tree visualization, a search program written as a nest of iterations instructs a computer to examine the tree in an orderly left-to-right manner as illustrated in Fig. 4.5. At each node we choose to examine next the branch initiated by the leftmost downward-emanating link which has not yet been followed. When at a terminal node on the bottom of the tree or at a node for which all downward emanating links have already been examined, we *backtrack*, i.e. we return to the node whence we came at the previous level, ready to follow the next (to the right) link emanating from that node. With respect to the nest, backtracking occurs when one iteration has been completed (or, perhaps, deemed unnecessary) and we wish to return to the previous iteration level and assign the next value to that loop index. We call this the *normal scan*. In general, information about each "active" node (a node lying on the path between the current position and the origin) must be saved in order that backtracking (passage to the next emanating link) may proceed smoothly. This is quite economical of storage as the number of active nodes cannot exceed the number of levels of the tree. [In the association of the use of nests and the use of recursive procedures (see § 1.5.4), the depth of nesting and depth of recursion correspond. Furthermore, variables which depend on the nest index in our approach correspond to variables that must be "stacked" (i.e. variable values are used in inverse order of their assignment) in the recursive procedure approach.]

The examination of all links leading downward from a particular node is handled notationally by the nest iteration (§ 1.5.1). Since this iteration must accommodate all nodes of the tree, it assumes a general form. The loop quantifier almost always depends

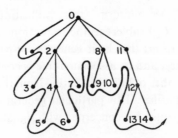

FIG. 4.5. Order of search for normal scan.

at least upon the nest index, and frequently upon more subtle parameters of the search or results computed during the search. The array of loop indices is referred to as the *nest vector*.

Many search programs take the generic form given in Fig. 4.6. The current values of the nest index i and the nest vector $j_{1 \text{ to } i}$ describe the progress of the scan through the

> INITIALIZE S_0, F
>
> with $i = 1$ to n:
>
> $\quad \cdot$
> $\quad \cdot$
>
> $\quad S_i = FUNCTION(i, S_{i-1}, j, F)$
>
> $\quad$ for $j_i \in S_i$:
>
> $\quad\quad$ CALCULATION OF $F(j_1, j_2, \ldots, j_i)$
>
> $\quad \cdot$
> $\quad \cdot$
>
> $\quad\quad$ USE $F(j_1, j_2, \ldots, j_n)$

FIG. 4.6. Generic form of nest for search programs.

tree. The possible values for a component of the nest vector, which determine the links emanating from a particular node, are given here as the set S_i; this set is often a function of i, previous "node sets", the vector j, and arbitrary functions computed during the scan. The objective of the tree scan is shown here as a function F, progressively computed as we move deeper into the nest; it is most usually a function of the (partial) nest vector, computed most efficiently as the search proceeds.

The time spent in scanning a search tree (i.e. in performing a backtrack calculation) is proportional to the number of nodes in the tree. The containment of this number to a reasonable size (search trees tend to be large) is the primary concern of the backtrack programmer. This subject is pursued later in this chapter and in § 6.6.

4.1.3. *Notational Variants*

Apart from straightforward nest resolution (Fig. 1.13) there are other ways of programming a tree scan without use of the with-statement structure. An important alternative, sometimes useful in generation procedures, is illustrated by segment A of Fig. 4.7. This program generates all *m*-component vectors whose components are natural numbers less than *n*. The equivalent program written as a nest of iterations is given as segment N in the figure.

It is useful to note the difference in viewpoints of these two equivalent programs. The nest program, segment N, emphasizes downward movement in the tree; backtracking (upward movement) occurs as a matter of course when an iteration is complete (i.e. when a component has exhausted its range). Segment A emphasizes backtracking by explicitly searching, beginning from a terminal node of the tree (line A.3), for the first

"Generation without nest"

A.1   for $i = 1$ to m: $j_i = 0$

A.2 $\#1$ USE $(j_1, j_2, \ldots, j_m)$

A.3   for $i = 1$ to m:

A.4 $j_i + 1 \rightarrow j_i$

A.5   if $j_i = n$: reiterate

A.6   for $k = 1$ to $i-1$: $j_k = 0$

A.7   go to $\#1$

"Generation with nest"

N.1   with $i = m, m-1, \ldots, 1$:

N.2 $\ddots$

N.3   for $j_i = 0$ to $n-1$:

N.4 $\ddots$

N.5   USE $(j_1, j_2, \ldots, j_m)$

FIG. 4.7. Backtrack generation without and with nest.

component which has not exhausted its range. At this point, previous components which had exhausted their range are reset to zero (line A.6). The less elegant, yet more elemental, structure of segment A has merit when algorithmic complications favor a reduction to simpler statements.

A different approach to tree examination is exemplified by the *list scan*. This method of organizing a search is presented not so much as an alternative effective in certain cases, but because it is a method which suggests itself quite naturally to many researchers; its relationship to backtracking should be understood. The idea is to examine, sequentially, nodes from a growing list. New nodes are added to the list as we follow the links emanating from the node currently being examined. If the tree is finite, the examination of nodes will sooner or later proceed faster than the growth of the list and eventually the list will be exhausted. The order of placement on the list is indicated by the node numbering illustrated in Fig. 4.8.

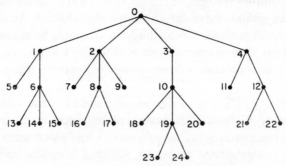

FIG. 4.8. Order of search for list scan.

$L_0 = $ INITIAL NODE; $I = 1$
for $i = 0, 1, \ldots, I-1$:
    for each LINK EMANATING FROM L_i, as $j = I, I+1, \ldots$:
      $L_j = $ NODE AT END OF LINK
      PROGRAM
    $I = j$

FIG. 4.9. Generic program structure for list scan.

A generic program for the list scan is given in Fig. 4.9. The list scan does not involve backtracking *per se* since no record of the tree structure is maintained, new nodes being added to a single list without reference to their tree level. As a rule, however, the list scan does require considerably more storage than the normal scan since on the average the current list is longer than the depth of the tree. Perhaps the most prominent attribute of a list scan is its simplicity; it can be useful in cases where indexing of tree levels and storage are not important.

Additional variants, perhaps more conceptual than notational, are treated later (notably in § 4.4). Notational variants suggested by precomputation, program specialization, etc. (§ 2.3), are, of course, ubiquitous. A certain amount of program experimentation and notational tailoring (to the system, tastes, and requirements of a particular researcher) is to be expected.

EXERCISES

1. The search tree of Fig. 4.1 represents the nine 2-digit base 3 numbers. Draw the search tree for the eight 3-digit base 2 numbers.

2. Draw the search tree associated with Fig. 4.2 with the nest quantifier replaced by "with $i = 3$ to 1:".

3. Draw the search tree associated with Fig. 4.4 with $n = 7$ and $m = 4$.

4. What program change is required to reverse the order in which the terminal nodes are examined, i.e. to examine a tree from right to left?

5. Given a positive integer n in prime factor representation, i.e. given $(p_1, p_2, \ldots, p_k)$ and $(e_1, e_2, \ldots, e_k)$, where

$$n = p_1^{e_1} p_2^{e_2} \ldots p_k^{e_k},$$

write a program for generating all factors of n in standard representation. Note that the precomputation of an array $G_{k,z} = p_k^z$ makes the generation more efficient.

6. (a) What signatures $(q_0, q_1, \ldots, q_{n-1})$ does the program of Fig. 4.10 generate?

(b) How does the search tree for this program (with $n = 4$) differ from that given in Fig. 4.3?

(c) Rewrite this program using a scalar S which serves the same function as the 1-array $S_{0 \text{ to } n}$.

$S_0 = $ (set) n
with $i = 0$ to $n-1$:
    $\cdot$
    $\cdot$
    for $q_i \in S_i$:
      $S_{i+1} = S_i \sim \{q_i\}$
      $\cdot$
      $\cdot$
    USE $(q_0, q_1, \ldots q_{n-1})$

FIG. 4.10. A common generation.

4.2. Basic Backtracking and Impasse Detection

The chief source of concern in constructing backtrack programs is the execution time of the resulting algorithm. A straightforward exhaustive search is often impracticable, even for a high-speed computer, due to the large number of nodes which must be individually examined (e.g. an n-level tree with M links emanating from each node has M^n terminal nodes). However, with ingenuity it is frequently possible to detect large branches of the tree which need not be followed at all, and a seemingly impossible search becomes feasible. We call this aspect of search design *impasse detection*.

Normal backtracking occurs when all links (hence all branches) emanating downward from a given node have been examined and we are ready to return (i.e. backtrack) to the previous node and follow the next link from there. Within a program normal backtracking is handled "automatically" by the index range specification given in the general iteration quantifier. Backtracking due to impasse detection occurs when a state is observed which obviates the investigation of the remaining branches emanating from a given node. It is indicated within a program in one of several ways, frequently using one of the auxiliary statements "reiterate", "exit from loop", or "backtrack" (see § 1.5.3). Reiteration (or delimitation of the range of the loop index by use of "such that" phrases—see § 1.4.4 and Fig. 4.11, for instance) occurs when a value for the loop index has been found sterile but other sibling values still need to be examined. The "backtrack" and "exit from loop" notations occur when the search has proceeded "up a blind alley", when the remaining values for the loop index at the present level can lead nowhere and the search should retreat up the tree.

It should be mentioned that the design of *effective* impasse detection calls for program experimentation as well as ingenuity. An actual bout with the computer can often save many hours of desk analysis. This section illustrates with specific examples certain general features of backtracking and impasse detection. Several genuine problems on which the reader himself may experiment are given in the exercises. (See Golomb and Baumert [111] for additional examples.)

4.2.1. *Graph Coloring*

Consider the problem of generating the different ways in which the vertices of a given graph may be colored using k colors labeled $0, 1, \ldots, k-1$ such that adjacent vertices have different colors (we call these *k-consistent colorings*, or *k-colorations*, of the graph). For purposes of this discussion we need not concern ourselves with isomorphic colorations produced either by symmetries of the graph or permutation of the colors. Thus we consider two colorations distinct if there exists a vertex of the graph which is assigned a different color under the two schemes. Let the graph G be given in vertex adjacency form, i.e. let G_i, $i = 0, 1, \ldots, n-1$, be the set of vertices adjacent to vertex i. Let $c_{0 \text{ to } n-1}$ be a 1-array whose entry c_i, $c_i \in k$, indicates the color assigned to vertex i. A straightforward but inefficient enumeration search is given as segment S in Fig. 4.11.

S.1   "Straightforward, but slow, program"

S.2   with $i = 0, 1, \ldots, n-1$:

S.3 $\ddots$

S.4   for $c_i = 0, 1, \ldots, k-1 \ni \forall_{x \in G_i \cap i}(c_x \neq c_i)$:

S.5 $\ddots$

S.6   USE COLORATION

I.1   "Improved program"

I.2   with $i = 0, 1, \ldots, n-1$:

I.3 $\ddots$

I.4 $C_i = k \sim \bigcup_{x \in G_i \cap i}(\{c_x\})$ "C_i is the set of colors"

I.5   for $c_i \in C_i$:   "available for vertex i"

I.6 $\ddots$

I.7   USE COLORATION

FIG. 4.11. Two straightforward graph coloring programs.

This program tests each possible color of vertex i against the color of adjacent, previously colored vertices. If a match exists, that color is disallowed and we test the next color. If no match exists, vertex i receives that color and we proceed to try to find a color for vertex $i+1$. Notationally, impasse detection appears in this example in the form of the rather stringent test (line S.4) which each acceptable component of the nest vector must pass.

Without altering the basic structure of this search, a considerable improvement may be attained by applying the principle of precomputation (§ 2.3.1). In segment I of Fig. 4.11, the "test" for illegality of the possible colors is made early, before a particular color is considered, and with a single examination of adjacent colored vertices. If, as in the slower program (segment S of Fig. 4.11), we delay testing for illegality until a particular color is considered, then the examination of adjacent colored vertices is repeated for each possible color. (Computer experiments with $k = 4$ and n about 30 indicate that the program of segment I requires only about 55% of the time required by the program of segment S.)

4.2.2. *Anticipated Impasse Detection*

The impasse which occurs when C_i of Fig. 4.11 is empty, i.e. when there are no available colors for vertex i, may be anticipated. Rather than compute the set of available colors for vertex i at the time a color is to be selected, it is more efficient to keep a running tally of the available colors for each vertex. In this way an impasse due to a "dead-end"

$$
\begin{array}{lll}
\text{C.1} & & \text{for } i;\, y:\, D_{i,y} = 0 \\
\text{C.2} & & \text{with } i = 0, 1, \ldots, n-1: \\
\text{C.3} & & \quad\ddots \\
\text{C.4} & & \quad \text{for } c_i = 0, 1, \ldots, k-1 \text{ such that } D_{i,c_i} = 0: \\
\text{C.5} & & \quad\quad \text{for } j \in G_i \sim i: \\
\text{C.6} & & \quad\quad\quad D_{j,c_i} + 1 \to D_{j,c_i} \\
\text{C.7} & & \quad\quad\quad \text{unless } \exists_{y=0 \text{ to } k-1}(D_{j,y} = 0): \\
\text{C.8} & & \quad\quad\quad\quad F = (G_i \sim i) \cap (j+1);\ \text{go to }\ \#1 \\
\text{C.9} & & \quad\quad\ddots \\
\text{C.10} & & \quad\quad\quad \text{USE COLORATION} \\
\text{C.11} & & \quad\quad F = G_i \sim i \\
\text{C.12} & \#1 & \quad\quad \text{for } j \in F: \\
\text{C.13} & & \quad\quad\quad D_{j,c_i} - 1 \to D_{j,c_i}
\end{array}
$$

Fig. 4.12. A coloring program using anticipated impasse detection.

vertex may be detected when it is created rather than possibly much later when that vertex is examined. The program of Fig. 4.12 is designed in this way.

The tally-keeping is accomplished here using a 2-array $D_{0 \text{ to } n-1,\, 0 \text{ to } k-1}$. An entry $D_{i,y}$ gives the number of neighbors of vertex i which have already been assigned color y. This tally is increased (at line C.6) for all uncolored neighbors of vertex i when color y is assigned to that vertex. When normal backtracking occurs, i.e. where we are about to try a new color for vertex i, the tally must be returned to its previous value with respect to the old color c_i. This is accomplished by the loop of lines C.11 to C.13. When we backtrack because of the detection of an impasse (i.e. the test of line C.7 is successful), the tally must be decreased for those neighbors for which it was just (prematurely) increased. This is accomplished by transferring control to the iteration of line C.12 after adjusting the index range F appropriately.

Anticipated impasse detection such as this prunes many unessential branches from the tree. However, such savings must be weighed against the cost of the impasse detection itself. If an involved test detects only few impasses or is successful only deep within a nest, it might add more time to a search than it saves. In the case at hand, the test of line C.7 is indeed involved but may be simplified by use of a better representation for D (see exercise 1)—an optimized version of this program has a running time roughly 75% of that for the better program of Fig. 4.11 (again with $k = 4$ and n about 30).

A particular form of anticipated impasse detection has been called "preclusion" by Golomb and Baumert [111]. In our example we could say the assignment of a color to a vertex "precludes" the assignment of that color to adjacent vertices. This viewpoint is useful notably in "covering problems" (§ 6.4), where the choice of a subset as part of the cover may preclude (or obviate) the choice of subsets which contain some of the same elements.

4.2.3. *Search Rearrangement*

A rule-of-thumb for backtracking is that, other things being equal, it is better to arrange a search so that nodes of low valence appear at shallow levels of the tree, or equivalently, so that iterations with the most restricted index ranges come early in the nest. [It is interesting to note that this is essentially the rule applied in 1823 by Warnsdorff to generate knight's tours (see Ball [2]). He suggested that a jump should be made to a square from which the tour had fewest possible continuations.] The application of this rule to general Hamiltonian path generation is discussed in § 4.2.4. With respect to impasse detection, such arrangement tends to merge many small unessential branches, each requiring discovery deep within the nest, into one large unessential branch detectable in one fell swoop at a shallow level of the nest. A search can sometimes be prearranged in

$$I_0 = 0; \ B = \text{(set)} \ n; \text{ for } h; \ x: D_{h,x} = 0$$
$$\text{with } h = 0, 1, \ldots, n-1:$$

$$.$$
$$.$$
$$.$$

$$i = I_h$$
$$B \sim \{i\} \to B$$
$$\text{for } c_i = 0, 1, \ldots, k-1 \ni D_{i,c_i} = 0:$$

$$M = k+1$$
$$\text{for } j \in B:$$

$$\text{if } j \in G_i:$$

$$D_{j,c_i} + 1 \to D_{j,c_i}$$
$$M^* = \sum_{y \in k \ni D_{j,y} \neq 0} (1)$$

$$\text{if } M^* = 0: F = G_i \cap B \cap (j+1); \text{ go to } \#1$$

$$\text{otherwise: } M^* = \sum_{y \in k \ni D_{j,y} \neq 0} (1)$$
$$\text{if } M^* < M: M = M^*; I_{h+1} = j$$

$$.$$
$$.$$
$$.$$

USE COLORATION

$$i = I_h$$

$$\#1 \qquad F = G_i \cap B$$

$$\text{for } j \in F:$$

$$D_{j,c_i} - 1 \to D_{j,c_i}$$

$$B \cup \{i\} \to B$$

FIG. 4.13. A coloring program using dynamic vertex rearrangement.

accordance with this rule, but often a dynamic adjustment is necessary or else more effective.

With regard to the graph coloring algorithms, this rule indicates that we should color next a vertex which has as few colors as possible available to it (of course, a backtrack is imminent if a vertex has no colors available to it). A prearrangement of the vertices so that the "next" vertex has the largest number of already labeled neighbors is in partial accord with this rule and is useful in the algorithms presented. An algorithm which dynamically selects a vertex with minimum available colors for the next vertex to be colored, hence which is in complete accord with the rule, is given in Fig. 4.13. In this program, the 1-array $I_{0 \text{ to } n-1}$ records the order in which the vertices are to be colored, i.e. I_h is the label of the $(h+1)^{\text{st}}$ vertex to be colored. The set B records the vertices yet to be colored, i.e. after h vertices have been colored, $B = n \sim \{I_0, I_1, \ldots, I_{h-1}\}$.

The sophisticated backtrack program of Fig. 4.13 is a long way from the straight-forward enumeration given in Fig. 4.11. Such sophistication of impasse detection and search rearrangement is absolutely essential for the successful completion of many combinatorial studies. (For the case at hand, however, the dynamic rearrangement of Fig. 4.13 turns out to be a negligible improvement over the program of Fig. 4.12 applied to carefully prearranged vertices.) On the other hand, straightforward and readable algorithms such as in Fig. 4.11 also have a place—for instance, when n is small or when the impasse detection is excessively involved, hence time-consuming. Indeed, the program-ming of a complex backtrack itself can involve extensive research in order to find the most suitable form for impasse detection and backtracking.

4.2.4. *Hamiltonian Path Generation*

A *Hamiltonian path* in a graph is a sequence of adjacent vertices (i.e. each vertex is adjacent in the graph to the preceding member of the sequence) which contains all the vertices of the graph once and only once. If the first member of the sequence is adjacent to the last member, then it is a *Hamiltonian cycle*. The edges formed by successive mem-bers of the path or cycle are also considered part of the path or cycle. Not all linear graphs possess Hamiltonian paths or cycles. In fact, since no generally applicable criteria for their existence in an arbitrary graph are known, their generation and/or enumeration involves an exhaustive search. As another example of impasse detection, we present, in Fig. 4.14, a program for generating all Hamiltonian paths in an arbitrary graph $G_{0 \text{ to } I-1}$ given in adjacency representation.

The idea is to do a backtrack generation of all paths in the graph using for impasse detection simple criteria applied to the successively smaller "section" subgraphs (see exercise 3 of § 3.4) obtained by deleting the vertices of the partial paths. The criteria used by this program are that a subgraph (1) must have no isolated vertices (unless the sub-graph consists of a single vertex), (2) must not contain more than two vertices with unit valence, and (3) must not contain two vertices with unit valence if neither of these vertices is taken as the next vertex of the path. That criterion (1) is necessary follows from the more

"Given: $G_{0 \text{ to } I-1}$ and I"

H.1 $M_1 = (\text{set})\ I;\ e_0 = 0$

H.2   if $impasse(M_1, 1) = 1$: go to #1

H.3   with $i = 1$ to $I-1$:

H.4 $\ddots$

H.5   for $v_i \in Q_i$; $j_i \in V_{i,v_i}$:

H.6   if $v_i = 1$: $e_i = e_{i-1} + |V_{i,1}| - 1$

H.7   otherwise: $e_i = e_{i-1} + |V_{i,1}|$

H.8 $M_{i+1} = M_i \sim \{j_i\}$

H.9   if $impasse(M_{i+1} \cap G_{j_i},\ i+1) = 1$: reiterate

H.10 $\ddots$

H.11   USE PATH $(j_1, j_2, \ldots, j_{I-1}, (M_I)_1)$

H.12   #1   GENERATION COMPLETE

I.0 $impasse(B, k)$

I.1 $(\text{set})\ B;\ Q_k = \{\ \}$

I.2   for $b \in B$:

I.3 $q = |M_k \cap G_b|$

I.4   if $q = 0$ and $k \neq I$: $impasse(\ ) = 1$; exit from procedure

I.5   if $q \notin Q_k$: $V_{k,q} = \{b\}$; $Q_k \cup \{q\} \to Q_k$

I.6   otherwise: $V_{k,q} \cup \{b\} \to V_{k,q}$

I.7 $d = e_{k-1} + |V_{k,1}|$

I.8   if $d > 2$ and $1 \in Q_k$: $impasse(\ ) = 1$; exit from procedure

I.9   if $d = 2$: $Q_k = \{1\}$

I.10 $impasse(\ ) = 0$; exit from procedure

Fig. 4.14. An efficient program for Hamiltonian path generation.

general statement that a "disconnected" graph (see § 6.3) cannot contain a Hamiltonian path. Criteria (2) and (3) point out the fact that vertices with unit valence can exist only as the initial and terminal vertices of a path.

Notationally, j_i is the ith vertex of the path being generated by the nest at lines H.3 to H.11 and M_i is the set of vertices of the ith section subgraph (from which j_i will be chosen) —M_1 contains all vertices of the graph. The tally e_i gives the number of vertices with unit valence in the subgraph specified by M_i. The co-procedure $impasse(\ )$ anticipates the failure of criteria (1) and (2) with respect to M_{i+1} (at lines I.4 and I.8 respectively). It also computes $V_{i,x}$, the set of vertices in M_i adjacent to j_{i-1} which have valence x, and Q_i, the set of such values of x. These quantities are used in the loop quantifier at line

H.5 to arrange the generation according to the rule-of-thumb given in § 4.2.3 (i.e. Warnsdorff's rule). Of course, the application of this rule is meaningless in this particular algorithm if the entire generation is carried out since *all* branches will eventually be tried. However, when existence of a Hamiltonian path is the sole reason for the search, the rule is effective since the *first* solution (if it exists) will be found much sooner.

Since *impasse*( ) is looking ahead, criterion (3) appears in a peculiar form. If the procedure discovers that exactly two possible next vertices of the path have only a single place to go (test of line I.9 is successful), then it makes sure that one of these is taken by restricting the possible values of v_{i+1} (the valence of the vertex considered for the next member of the path) to unity.

Note that the nest only produces the first $I-1$ members of the path, the final member is given simply by $(M_I)_1$, certain to be adjacent to j_{I-1} because of the repeated fulfillment of criterion (1). This notational trick, which shortens the running time of the program slightly, is common in backtrack programming. We leave it to the reader to discover other minor notational variations which make this algorithm more efficient.

EXERCISES

1. (a) Modify the program of Fig. 4.12 to use a set D_i^* which yields the number of neighbors of vertex i which have been assigned color y by the expression

$$|D_i^* \cap \{y, y+k, y+2k, \ldots, y+J \times K\}|$$

(*J* a sufficiently large constant) instead of by the value of the real variable $D_{i,y}$. (Hint: It is helpful to precompute a 2-array Q, $Q_{y,0} = \{ \}$ and $Q_{y,j} = \bigcup_{i=0 \text{ to } j-1}(\{y+ik\})$ for $j = 1$ to J.)

(b) Modify the program of Fig. 4.13 using this representation.

*2. The graph coloration generation schemes discussed here are inefficient when one wishes only to *count* colorations or merely to test for their existence. An effective counting algorithm is given by Read [128]. Let $G_{0 \text{ to } v-1}$ be an arbitrary graph. If G is complete, then the number of k-colorations of G, $C(G)$, is $k(k-1) \ldots (k-v+1)$. Otherwise, there exist a pair $\{i,j\}$ which is not an edge of G. Let $G1$ be the graph obtained from G by including that pair as an edge and let $G2$ be the graph obtained by equating i with j (eliminating multiple edges). Then $C(G) = C(G1) + C(G2)$ by the rule-of-sum. Write a "branching program" (i.e. backtrack program with binary search tree) for this method.

*3. A *knight's tour* is a Hamiltonian path on the graph constructed from an $m \times n$ chessboard with two squares adjacent if the chesspiece "knight" can jump in a single move between the squares (see § 3.6). Make a table showing the number of knight's tours for $m, n = 3, 4, 5, 6$. (Do not consider symmetry, i.e. with a fixed labeling of the squares, regard two tours as distinct if the sequence of labels defining the paths is distinct.)

*4. Write a program to discover if the vertices of a given graph may be labeled in such a way that the absolute value of the difference of the labels attached to adjacent vertices does not exceed k.

*5. An *arc* in a graph is a path in which every vertex is traversed (i.e. appears as second component and then first component in successive edges of the path) at most once. An arc *connects* a vertex appearing in one of its edges with each vertex appearing in a succeeding edge of the arc. The *distance* between two vertices of a graph is the number of edges in the shortest arc connecting the two vertices. Write a program to calculate the length (i.e. number of edges) of the longest arc in the n-cube, subject to the restriction that the distance between each two vertices of the arc, measured in the subgraph obtained by deleting the edges of the arc from the n-cube, is at least k. (This is a generalization of the snake-in-the-box problem—see [106], for instance.)

6. How are the Hamiltonian path impasse criteria of § 4.2.4 affected when Hamiltonian *cycles* are being generated? Write a program to generate Hamiltonian cycles. (See also exercise 11 of § 6.3.)

*7. A graph is *planar* if it can be drawn in the plane with edges intersecting only at vertices. A planar graph is *maximal* if no additional pair of existing vertices may be included as an edge of the graph without destroying the planarity of the graph ("faces" of the graph are thus triangles). A theorem of Heawood (see chapter 9 of Ore [235], for instance) says that a maximal planar graph may be consistently colored in four colors if and only if the numbers $+1$ and -1 may be assigned to the triangular faces of the graph in such a way that the sum of numbers assigned to faces "adjacent" to each vertex is congruent to zero modulo 3.

(a) Write a program for testing the four-colorability of an arbitrary given maximal planar graph using this criterion (in this respect see Yamabe and Pope [134]).

(b) Using Euler's formula for polyhedra—i.e. number of vertices minus number of edges plus number of faces equals two—compare the search tree hence the effectiveness of this approach to that of the more direct approach used in the text.

*8. A Hamiltonian path on the N-cube is equivalent to an N-bit Gray code. Find an N-bit Gray code for $4 \leqslant N \leqslant 9$ which is not the code described in § 3.1.2.

*9. A comma-free code is a collection of words, a dictionary, from which "sentences" which do not contain spaces or other punctuation may be formed without the slightest ambiguity. For instance, if "good" and "dog" are in the dictionary, then "odd" could not be since the sentence "gooddog" contains an overlap ambiguity. Consider dictionaries containing only n letter words, each word formed from a k letter alphabet, and let $W(k,n)$ be the greatest number of words that such a dictionary could contain. Show that $W(4,4) = 57$. Is the calculation of $W(6,3)$ feasible? (See Golomb [110].)

4.3. Optimization Backtracking

Backtrack programming is sometimes useful, or necessary, in certain problems of optimization, problems which involve search for an "optimum" member of a finite class of objects whose generation has an arborescent structure. In these cases, the "appraisal calculation" is merged with the generation and used during this tree scan to eliminate branches which cannot produce an object better than the best already generated. In the literature this technique is often called "branch and bound" [120] or "truncated enumeration" [109], as well as backtrack programming.

For terminology, let P be the class of objects, the *object-space*, and $app(p)$, $p \in P$, be a given real valued function with domain P called the *appraisal (function)*. The goal of the search is to find an element of P, say q, such that $app(q)$ is optimum (e.g. maximum). The objects of P are generated in a systematic tree-like manner; the appraisal is calculated as part of the generation and is used for impasse detection during the tree scan.

4.3.1. *Branch and Bound—The Traveling Salesman Problem*

We use a particular problem, the traveling salesman problem (abbreviated to TSP), to illustrate branch and bound methods.

The TSP consists of finding the cheapest route a salesman can follow in visiting various cities given the cost of traveling between any two cities. Let the salesman's home city be labeled 0 and the outlying cities $1, 2, \ldots, n-1$. We are given the expense $E_{i,j}$, i and

$j = 0$ to $n-1$, of traveling from city i to city j ($E_{i,j}$ is not necessarily equal to $E_{j,i}$) and wish to minimize

$$app(q) = \sum_{i=1 \text{ to } n}(E_{q_{i-1}, q_i}), \tag{4.1}$$

where $q_0 = q_n = 0$ and $(q_1, q_2, \ldots, q_{n-1})$ is a permutation of $1, 2, \ldots, n-1$, q_i giving the ith city to be visited on the n link cyclic tour. (An alternate formulation is given in § 4.3.2.)

Observe that the notation

$$MIN_{route(n,q)} \left(\sum_i (E_{q_{i-1}, q_i})\right),$$

where *route*( ) is a generation procedure (§ 1.4.2) forming in turn each possible route, would indeed produce the length of the minimum route (given enough time). However, the backtrack program (the route generation procedure) is divorced from the appraisal evaluation (the summation of the expenses). It is the merging of these two jobs which permits effective optimization impasse detection.

The basic backtrack program for solving the TSP is the generation of the possible routes—in this formulation, the generation of the permutations of the $n-1$ marks 1, 2, ..., $n-1$. Using the structure of the lexicographic generation given in § 5.2.1 (and exercise 6 of § 4.1) a crude but illustrative program for solving the TSP is given in Fig. 4.15. The final appraisal of the route from city 0 to city q_1 to ... to city q_{n-1} and back to city 0 is given by *App*. Partial appraisals (i.e. partial sums) are given by $A_i, i = 0, 1, \ldots, n-1$. The variable T gives the smallest appraisal so far calculated. It is initially set to T^* which either is larger than any possible route cost or is an upper limit for appraisals of interesting routes.

The optimization impasse detection occurs at line T.7. This test asks if the route given by $q_1, q_2, \ldots, q_i$ completed by $n-i+1$ trips between the two "closest" cities, each with cost K (line T.1), is more expensive than the best route so far generated. While eliminating the scan of many branches of the search tree, this test is actually fairly weak, allowing

T.1	$T = T^*$; $K = MIN_{i;j}(E_{i,j})$
T.2	$S_1 = (n-1) \oplus \{1\}$; $A_0 = 0$; $q_0 = 0$
T.3	with $i = 1$ to $n-1$:
T.4	$\ddots$
T.5	for $q_i \in S_i$:
T.6	$A_i = A_{i-1} + E_{q_{i-1}, q_i}$
T.7	if $A_i + (n-i+1)K > T$: reiterate
T.8	$S_{i+1} = S_i \sim \{q_i\}$
T.9	$\ddots$
T.10	$App = A_{n-1} + E_{q_{n-1}, 0}$
T.11	if $App < T$: $T = App$; $t = q$

FIG. 4.15. An unsophisticated TSP algorithm.

rather deep penetration into the nest before an impasse is discovered. In general, more elaborate optimization impasse detection, i.e. computation of a better bound for the appraisal function, allows backtracking at a shallower level of the tree. Arriving at a balance between sophistication of impasse detection and extent of tree scan which yields an effective algorithm for an optimization problem often requires considerable "on-line" experimentation since backtrack programs are difficult to analyze at one's desk.

Besides its modest impasse detection, an important shortcoming of this algorithm is the fixed and arbitrary ordering of the cities. If city 1 is a long way from city 0, then much time will be wasted in scanning the part of the tree based on traveling from city 0 to city 1, a link which is probably not part of the optimal tour. A general principle of optimization backtracking (consistent with the rule-of-thumb discussed in § 4.2.3) is that it is highly desirable to find good answers early in the search, for then the generation of many poor answers will be entirely omitted.

4.3.2. *The Algorithm of Little, Murty, Sweeney, and Karel*

A branch and bound algorithm for the TSP, which effectively attacks both of the shortcomings of the naïve Fig. 4.15 program, has been given by Little *et al.* [124]. For this discussion we view the TSP as a network problem with $E_{\_,\_}$ as the network incidence matrix. A solution is then a set of n network edges (I_k, J_k), $k = 0$ to $n-1$, which form a Hamiltonian cycle and for which

$$S = \sum_k (E_{I_k, J_k})$$

is a minimum.

Since a solution specifies an entry in each row of the expense matrix, we have

$$S \geqslant \sum_{i \in n} (m_i)$$

where

$$m_i = MIN_{j \in n \ni j \neq i} (E_{i,j}).$$

Furthermore, each minimum m_i may be subtracted from every entry in its row without affecting the solution set for the problem. By the same token we have further that

$$S \geqslant \sum_{i \in n} (m_i) + \sum_{j \in n} (m_j^*) \tag{4.2}$$

where

$$m_j^* = MIN_{i \in n \ni i \neq j} (E_{i,j} - m_i);$$

and the expense matrix $E_{i,j} - m_i - m_j^*$, $i \in n$, $j \in n$, produces the same solution cycles as does $E_{\_,\_}$. Note that this reduced matrix has at least one zero in every row and every column.

Now consider the following branching process and resulting binary search tree (characteristic of many branch and bound algorithms). Each node of the tree corresponds to a TSP with its network reduced as discussed above—the origin corresponds to the original reduced TSP. The edges of the corresponding network are classified as "available" or "unavailable". An edge is *unavailable* (for use in the cyclic tour) if a cycle with too few edges would be formed or if that edge is outlawed during the branching process—see (2) below. (The edges (i,i), $i \in n$, at least, are unavailable in the original problem.) For each problem an available zero of the expense matrix is chosen (by a criterion to be discussed shortly) and the following two exhaustive and mutually exclusive possibilities are considered: (1) the edge of the network corresponding to the zero is *included in* the cyclic tour, and (2) that edge is *outlawed from* the tour. Possibility (1) leads to a TSP with one less city, while possibility (2) leads in effect to a TSP with one more unavailable edge.

In the binary search tree we imagine possibility (1) as the left link and possibility (2) as the right link. When n left links have been taken, a trial solution to the original problem

L.1 $r, c = (\text{set}) \, n$; $A_0 = 0$; $T = T^*$; $X_{0,-,-} = E_{-,-}$

L.2   for $i \in n$: $a_i, a_i^* = (\text{set}) \, n \sim \{i\}$; $Zr_{0,i}, Zc_{0,i} = (\text{set}) \, n$; $p_i, p_i^* = n$

L.3   with $k = 1, 2, \ldots$, until $|r| = 1$:

L.4 $\ddots$

L.5 $A_k = A_{k-1}$; for $j \in c$: $Zc_{k,j} = Zc_{k-1,j}$

L.6   for $i \in r$: $Zr_{k,i} = Zr_{k-1,i}$; for $j \in c$: $X_{k,i,j} = X_{k-1,i,j}$

L.7   for $i \in r \ni c \cap a_i \subset Zr_{k,i}$:[1]

L.8 $m = MIN_{j \in c \cap a_i}(X_{k,i,j})$; $A_k + m \rightarrow A_k$

L.9   for $j \in c \cap a_i$:

L.10 $X_{k,i,j} - m \rightarrow X_{k,i,j}, x$

L.11   if $x = 0$: $Zr_{k,i} \sim \{j\} \rightarrow Zr_{k,i}$; $Zc_{k,j} \sim \{i\} \rightarrow Zc_{k,j}$

L.12   for $j \in c \ni r \cap a_j^* \subset Zc_{k,j}$:

L.13 $m = MIN_{i \in r \cap a_j^*}(X_{k,i,j})$; $A_k + m \rightarrow A_k$

L.14   for $i \in r \cap a_j^*$:

L.15 $X_{k,i,j} - m \rightarrow X_{k,i,j}, x$

L.16   if $x = 0$: $Zr_{k,i} \sim \{j\} \rightarrow Zr_{k,i}$; $Zc_{k,j} \sim \{i\} \rightarrow Zc_{k,j}$

L.17 $M = -1$

L.18   for $i \in r$; $j \in (c \sim Zr_{k,i}) \cap a_i$:

L.19 $m = MIN_{h \in (c \cap a_i) \sim \{j\}}(X_{k,i,h})$

L.20 $m^* = MIN_{h \in (r \cap a_j^*) \sim \{i\}}(X_{k,h,j})$

L.21   if $m + m^* > M$: $I_k = i$; $J_k = j$; $M = m + m^*$

L.22   if $M \geqslant 0$ and $A_k < T$:

L.23   for *inandout*( ):

L.24 $\ddots$

L.25 $t = p$; $p_{(r)_1} = (c)_1$; $T = A_{k-1}$

FIG. 4.16A. The basic generation for the algorithm of Little *et al.*

has been attained. The tree is finite since eventually (after at most $n(n-1)$ right links have been taken) there are no available edges upon which to branch and backtracking may be effected. Moreover, regardless of which edges are used for branching, every Hamiltonian cycle appears as a terminal node somewhere in the tree (the reader should convince himself of this fact).

The algorithm consists of the merging of the appraisal evaluation (4.2) and associated matrix reduction with this arborescent cycle generation along with a rule for selecting the network edge upon which to *branch*. The rule suggested by Little *et al.*, which turns out to be extremely effective (i.e. good answers are produced early), is to choose a zero yielding the largest (i.e. worst) appraisal for the problem of possibility (2), i.e. for the node at the end of the right link.

The complete algorithm is implemented in Figs. 4.16A and B. The basic generation appears as the nest in Fig. 4.16A; the bookkeeping for the branching process is handled by the *inandout*( ) generation procedure of Fig. 4.16B. We now let k be the nest index, but, as in Fig. 4.15, E is the expense matrix, A is the vector of partial appraisals, and T is the smallest appraisal so far calculated. Here the partial appraisal A_k also serves as the lower bound on which the optimization impasse detection is based (line L.22).

Due to the matrix reductions, entries of the original expense matrix are changed at the various nodes of the search tree. Thus an array $X_{k,-,-}$, $k = 0, 1, \ldots$, of expense matrices is needed. Also, it is convenient to have at one's disposal various auxiliary masks

L.B.0   *inandout*( )

L.B.1 $i = I_k; j = J_k$

L.B.2   if *Entry* $= 1$:

L.B.3 $r \sim \{i\} \to r; c \sim \{j\} \to c; p_i = j; p_j^* = i$

L.B.4   until $p_j = n: p_j \to j$

L.B.5   until $p_i^* = n: p_i^* \to i$

L.B.6 $a_j \sim \{i\} \to a_j; a_i^* \sim \{j\} \to a_i^*$

L.B.7 $I_k^* = i; J_k^* = j$

L.B.8   otherwise:

L.B.9   if $i \in r$:

L.B.10 $a_i \cup \{j\} \to a_i; a_j^* \cup \{i\} \to a_j^*$

L.B.11   *Exit* $= 1$; exit from procedure

L.B.12 $r \cup \{i\} \to r; c \cup \{j\} \to c$

L.B.13 $a_{J_k^*} \cup \{I_k^*\} \to a_{J_k^*}; a_{I_k^*}^* \cup \{J_k^*\} \to a_{I_k^*}^*$

L.B.14 $p_i, p_j^* = n$

L.B.15 $a_i \sim \{j\} \to a_i; a_j^* \sim \{i\} \to a_j^*$

L.B.16   *Exit* $= 0$; exit from procedure

FIG. 4.16B. The branching procedure for the algorithm of Little *et al.*

which describe assorted features of these expense matrices. Thus (1) r and c, which are subsets of $\{0,1,\ldots,n-1\}$, indicate respectively the rows and columns of $X_{k,-,-}$, (2) a_i, $i \in r$, and a_j^*, $j \in c$, indicate respectively the columns in row i and rows in column j which are still available for incorporation in the cycle, and (3) $Zr_{k,i}$, $i \in r$, and $Zc_{k,j}$, $j \in c$, indicate respectively the nonzero entries of row i and column j of $X_{k,-,-}$. The edges chosen for branching are given by (I_k, J_k), $k = 1, 2, \ldots$. The tour itself is given by $(p_0, p_1, \ldots, p_{n-1})$, where p_i indicates the city following city i on the tour. It is convenient also to have available $p_{0 \text{ to } n-1}^*$, the permutation inverse to $p_{0 \text{ to } n-1}$. We let $p_i = n$ when a successor city to city i has yet to be established; analogously, $p_j^* = n$ when no predecessor to city j exists.

The reduction of the matrices occurs at lines L.7 through L.16. The search for the proper branching edge occurs at lines L.17 through L.21. In this loop, $m + m^*$ gives the increase in the appraisal provided edge (i,j) is made unavailable; M records the largest such increase. If an edge for branching does exist and if our appraisal is still not too big (line L.22), then branching is initiated (line L.23). The left branch is handled by lines L.B.2 through L.B.7 of the *inandout*( ) procedure. The r and c masks are adjusted and the tour permutations extended at line B.3. At lines B.4 and B.5, a scan is made to determine what edge should be made unavailable (at line B.6) by virtue of completing a cycle with fewer than n edges. (Note, in line L.3, that the nest is terminated one step before the nth edge of the cycle would be outlawed.) This edge is recorded at line B.7 as (I_k^*, J_k^*) so that it can later, at line B.13, be reinstated as available. The right branch is handled at lines B.9 through B.15. Lines B.12 through B.14 undo the effect of having taken the left branch, while line B.15 makes the chosen (line L.21) edge unavailable. When the test at line B.9 is successful, both branches have been taken. We then undo the effect of having taken the right branch (line B.10) and prepare to backtrack.

As always, many notational variations to this particular program are possible. For instance, the second condition of line L.22 may be tested *prior* to the search of lines L.17 through L.21. We leave the matter of notational experimentation mostly to the reader. However, there is one important notational aspect which should be mentioned. During the execution of this program, the index sets for the minimization operations (e.g., $c \cap a_i$ at line L.8) at times become empty. It is convenient to have $MIN_{x \in \{ \}}(\text{EXPRESSION})$ defined as $+\infty$ (see exercise 9 of § 1.4, and § 2.1.4), or at least as some very large integer, for then the appraisal calculation at lines L.8 and L.13 and the test at line L.22 are treated consistently. However, for most computer systems, additional statements will have to be inserted into the program to accommodate these special (but not unusual) cases.

An excellent general survey of branch and bound methods is given by Lawler and Wood [120]. Further discussion of branching processes appears here in § 6.4. Gomory [112] gives a good general survey of the TSP. Modification of the algorithm presented here for different optimization problems (e.g. the *assignment problem*) is pursued in the exercises.

4.3.3. *Min–Max Optimization—Computer Chess*

Slightly more involved optimization backtracking occurs in programs for playing "two-person" games such as chess, go, and tick-tack-toe. These games may be represented by a (search) tree of variations in which the nodes correspond to "positions" of the game and links correspond to legal "moves". If the game is finite, then terminal nodes correspond to positions in which a decision has been reached—i.e. "white wins", "black wins", or "draw". Each level of the tree contains nodes corresponding to positions in which one of the players is "on the move". For chess, alternate levels give positions in which it is white's (or black's) move. A simple tree is given in Fig. 4.17. The numbers $+1$, -1, and 0 attached to the terminal nodes indicate that white wins, black wins, or the result is a draw respectively. With these outcome evaluations, and assuming intelligent play (i.e. each player is trying to win, or at least not lose), an evaluation has been given to the nonterminal positions of the game. Proceeding from the terminal nodes, the evaluation assigned to a node is the minimum (respectively, maximum) of the evaluations for its successor nodes when black (respectively, white) is to move. Analysis shows that in this game white has a "forced win", i.e. there exists a first move for white such that no matter what first move black makes there exists a second move for white such that no matter what second move black makes, white wins.

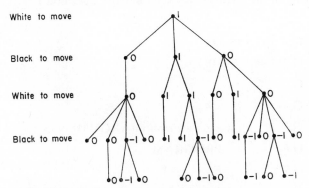

Fig. 4.17. A sample game tree and evaluation.

Given a position, a (machine or human) player bases a decision of which move to make upon an examination of the search tree. Of course, except in very special cases —e.g. simple chess problems and tick-tack-toe—one (even a computer) cannot expect to have terminal nodes, with their definite evaluation, within reach. However, a subjective evaluation of the position at a pseudo-terminal node can be made and the analysis can proceed as before.

Backtrack programming arises in the examination of a branch of the tree of variations. Let it be white's move. We restrict ourselves to $2n$ move analysis, i.e. to variations containing n moves by white and n response by black. Let E_i, $i = 0, 1, \ldots, 2n$, be the evaluation of a position at level i, $-1 \leqslant E_i \leqslant +1$. Thus E_{2n} assumes in turn each of the pseudo-evaluations assigned to the positions occurring at the end of the variations. For $i = 0, 1, \ldots, 2n-1$

$$E_i = \begin{cases} MAX_{\text{white moves}} (E_{i+1}), & \text{if } i \text{ is even} \\ MIN_{\text{black moves}} (E_{i+1}), & \text{if } i \text{ is odd.} \end{cases}$$

The object of the tree analysis is to choose a "best" move available to white from the given position, that is, to choose a move from which E_0 was derived. The program skeleton given in Fig. 4.18 outlines a backtrack search for accomplishing this analysis. (For notational convenience we have introduced the constant $E_{-1} = +1$.)

The optimization impasse detection occurs here at lines G.13 and G.15. As with the test at line T.7 in Fig. 4.15, many branches (i.e. variations) are omitted from examination when these tests are successful. In this case, however, we discover the opportunity for backtracking at a deeper level, hence the "exit from loop" in place of the "reiterate". This is due to the "minimax" nature of the optimization. The essence of the test at line G.13 is that if for a particular white move, a powerful black response is discovered, then we do not have to consider other black responses, or variations derived from them, as we have already shown the ineptness of the contemplated white move—we may proceed immediately with the next white move. (The term "$\alpha - \beta$ heuristic", due to McCarthy —see [132]—occurs in the literature for such impasse detection.)

G.1 $E_{-1} = +1$

G.2   with $i = 0, 2, \ldots, 2n-2$:

G.3 $\ddots$

G.4 $E_i = -1$

G.5   for each *whitemove*(ARGUMENTS):

G.6   PROGRAM

G.7 $E_{i+1} = +1$

G.8   for each *blackmove*(ARGUMENTS):

G.9   PROGRAM

G.10 $\ddots$

G.11   COMPUTE E_{2n}

G.12 $E_{i+1} = min(E_{i+2}, E_{i+1})$

G.13   if $E_{i+1} < E_i$: exit from loop

G.14 $E_i = max(E_i, E_{i+1})$

G.15   if $E_i > E_{i-1}$: exit from loop

FIG. 4.18. Evaluation bookkeeping for chess playing.

4.3.4. *Local Optima by Method-of-ascent*

In order to find and verify a "globally" optimal solution to a finite problem, such as the TSP, one must exhaustively scan the entire object-space P, either by a backtrack program as illustrated in Fig. 4.16 or by a "branch-merged tree search" (discussed in § 4.4). Even with extremely clever impasse detection such an exhaustive search can be impracticable in large problems. However, an intelligent partial scan of the object-space yielding a near-optimal solution is often possible.

One approach to finding approximate solutions is to locate "local optima" by means of the *method-of-ascent*. This method is applicable when there exists a distance function on P such that $|app(p) - app(q)|$ is relatively small when the distance from p to q is small. Starting at an arbitrary point of the object-space, a base-point, one moves to a neighboring point of the object-space which has a better appraisal. By a succession of such moves (changes of base-point) one eventually arrives at a *local optimum*—a point none of whose neighbors has better appraisal.

For most optimization problems a natural distance function exists—n minus the number of edges which two Hamiltonian cycles have in common is a natural distance function for the object-space of the TSP, for instance. In the simplest cases of local optimum determination (see § 4.3.5 for generalization), one treats points whose distance is minimum as neighboring points. In the unsymmetric TSP ($E_{i,j} \neq E_{j,i}$), two cycles containing $n-3$ common edges are neighbors. We present in Fig. 4.19 a straightforward method-of-ascent algorithm for the unsymmetric TSP. We are given a Hamiltonian cycle which represents a sub-optimal solution and wish to "converge" to a local minimum. The cycle, which might have been constructed from a partial execution of the algorithm of Fig. 4.16 (or perhaps at random), is given as the vector $(q_0, q_1, \ldots, q_{n-1})$, where q_i gives the ith city to be visited. (Note that this is the permutation of (4.1) and Fig. 4.15 but is not the same as the permutation $p_{0 \text{ to } n-1}$ of Fig. 4.16.) A neighboring cycle is formed by interchanging two consecutive cities of the tour; this changes exactly three edges of the cycle. The expense matrix E is as discussed previously.

"Method-of-ascent for TSP"

$b = 0; c = 1; d = 2$ "$n \geqslant 3$"

\# 1 for $j = 1$ to n, as $a = b, \ldots, b \rightarrow a, \ldots$, as $b = c, \ldots, c \rightarrow b, \ldots$,

as $c = d, \ldots, d \rightarrow c, \ldots$, as $d = d+1 \pmod n, \ldots, d+1 \pmod n \rightarrow d, \ldots$:

if $E_{q_a,q_c} + E_{q_c,q_b} + E_{q_b,q_d} < E_{q_a,q_b} + E_{q_b,q_c} + E_{q_c,q_d}$:

$q_b \leftrightarrow q_c$; go to \# 1

"At this point, a local minimum is achieved."

FIG. 4.19. Method-of-ascent applied to TSP.

Besides choosing any neighbor which has a better appraisal (as in the example of Fig. 4.19), one may appraise all neighbors and choose the neighbor with the best appraisal as the new base-point. Doing this will probably achieve a local optimum in fewer steps but at a significantly higher cost for each step. On the other hand, one can argue (see Gleason [108]) that one is apt to end up at a better local optimum if one chooses for the next base-point the neighbor yielding the least improvement over the current point. However, for most applications it is probably better to ignore both of these two approaches—the straightforward method-of-ascent exemplified in Fig. 4.19 is quite satisfactory. The case where one must find *all* optima accessible from a given position by ascent, without intermediate descent, is pursued in exercise 10.

4.3.5. *More Refined Approximation*

The local optimum search just described moves about the object-space in *minimal* steps, eventually locating a point none of whose *nearest* neighbors has a better appraisal. Instead of testing only minimal-distance neighbors, one may test (with considerably more effort, of course) points farther from the base-point, arriving eventually at a local optimum possibly superior to one obtained by the minimal-distance search. In fact, let A be an appraisal function defined on an object-set P and let d be a distance function which assumes only the values $D_1 < D_2 < \ldots < D_I$ (i.e. $d(p,q) \leqslant D_I$). Let

$$Q_i = \bigcup_{p \in P \ni \forall_{q \in P \ni d(p,q) \leqslant D_i} [A(q) \leqslant A(p)]} (\{p\}).$$

In other words, Q_i is the set of local optima using as neighbors to a base-point all points with distance $\leqslant D_i$. We have that

$$Q_1 \supset Q_2 \supset \ldots \supset Q_I$$

and the points in Q_I represent *globally* optimal solutions to a given problem.

Verification that a particular point x is a member of Q_i becomes increasingly more difficult for large i, and, in fact, verification that $x \in Q_I$ is tantamount to an exhaustive scan of P. It is noticed in practice, however, that points in Q_i for even small i have a nonnegligible probability of being optimal and can often serve as satisfactory approximate solutions to a problem. Lin [123] bases his effective TSP algorithm on finding, and comparing, members of Q_2. The author of this book has shown empirically [253] that this approach is useful for Boolean expression minimization.

Figure 4.20 outlines the program structure for this approach to optimization as it frequently occurs in practice. The procedure *neighbor*(p_{j-1}) generates minimal-distance neighbors of the point p_{j-1}; points further removed from the base-point p_0 are reached by a succession of these short steps. If a point is found which has an appraisal better than the current base-point, this point becomes the new base-point and calculation begins anew (line A.8). Note that this program structure is a direct generalization of that for finding a local optimum (see Fig. 4.19). (This structure is also closely related to the cut-off scan—see Fig. 4.22 in § 4.4.1.)

A.1 $p_0 =$ ARBITRARY POINT IN OBJECT SPACE

A.2   #1   for $n = 1$ to N: "N is the 'degree' of approximation"

A.3   with $j = 1$ to n:

A.4 $\ddots$

A.5   for each $neighbor(p_{j-1}) \to p_j$:

A.6 $\ddots$

A.7   if $a(p_n) < a(p_0)$:

A.8 $p_0 = p_n$; go to #1

A.9   "p_0 is 'N-optimal'."

FIG. 4.20. Structure of approximation algorithms.

This method suffers significantly from extensive repetition of computation, particularly for large N, since no record is kept of the region in the object-space already scanned. In general the elimination of this redundant computation is a difficult problem; one must usually exploit peculiar properties of a particular object-space to improve the situation measurably. We do not pursue this subject further, but do attempt in the exercises to familiarize the reader with various particular object-spaces.

EXERCISES

1. Construct a TSP algorithm based on the Hamiltonian cycle generation discussed in § 4.2.4 (and exercise 6 of § 4.2). Arrange the algorithm so that travel to the nearest unvisited city is attempted first. For optimization impasse detection, use as a lower bound the length of the path so far generated plus the minimum sum of distances among unvisited cities without regard to cycle formation.

2. (a) Prove that every Hamiltonian cycle for an arbitrary network would be generated by the branching process of § 4.3.2 (without optimization impasse detection, of course).

(b) Write a Hamiltonian cycle (of a graph) generation program based on the branching process of § 4.3.2.

3. Draw the search tree associated with the execution of the algorithm of Little *et al.*, as applied to the network

	0	1	2	3	4	5
0	╳	62	37	30	9	30
1	44	╳	10	97	77	20
2	23	78	╳	2	29	85
3	55	11	57	╳	93	41
4	69	95	50	89	╳	39
5	30	27	74	93	23	╳

4. The "first" complete cyclic tour generated by the algorithm of Little *et al.* (i.e. the cycle specified by p the first time the center of the nest is reached) is very often an optimal solution to the given TSP. Construct a network for which the first solution is not optimal.

5. The *assignment problem* consists of finding the permutation $(p_0, p_1, \ldots, p_{n-1})$ which minimizes

$$\sum_i (C_{i,p_i})$$

for a given cost matrix $C_{0\ to\ n-1,\ 0\ to\ n-1}$. Modify the algorithm of Fig. 4.16 to solve the assignment problem. (Hint: The modification is a simplification. This approach is closely related to the so-called *Hungarian method* due to Kuhn [118]—see Munkres [126], for instance. Also, see § 7.5.)

6. Show that with proper play on both sides, a game of tick-tack-toe always ends in a draw. (Hint: Draw the tree of variations, making use of the symmetries of the board—see § 6.5.2.)

7. Rewrite the program of Fig. 4.18 so that the nest contains a *single* iteration which handles moves of either player. Generalize the notation and program structure to handle an *n*-player game.

8. How does the general principle of optimization backtracking that "good answers should be found early in the search" relate to computer chess?

9. Write a program to select N points $x_{i_1}, x_{i_2}, \ldots, x_{i_N}$ from M points $x_1, x_2, \ldots, x_M$ which maximizes

$$\sum_{j=1\ to\ N-1}(|fct(x_{i_j})-fct(x_{i_{j+1}})|)$$

subject to the restriction $fct(x_{i_j})\cdot fct(x_{i_{j+1}}) < 0$, where $fct(\ )$ is an arbitrary real-valued function defined on the points. See Lynch [125] for a recursive procedure solution to this problem.

*10. Write a generic backtrack program for locating *all* local optima accessible from a given point of an object-space via a nondecreasing path of nearest neighbors. Since plateaus may exist, you must incorporate into your program a means to prevent cyclic wandering.

*11. Let M be an arbitrary $m\times m$ matrix of zeros and ones. For each of the 2^{2m} subsets of rows and columns, there corresponds a new matrix formed by interchanging zeros with ones in each of the rows and columns of the subset (an entry in a specified row *and* specified column thus remains unchanged).

(a) Write a program to find a matrix equivalent to a given matrix under a succession of these operations which contains the maximal number of ones.

(b) What degree of approximation—see Fig. 4.20—is necessary to insure that an optimal solution is obtained about one-quarter of the time for $m = 10$? (Note that the object-space is the $2m$-cube where adjacent vertices are neighbors. See Gleason [108].)

12. (a) How does the algorithm of Fig. 4.19 at least partially solve the problem of unnecessary repeated scan of object-space points?

(b) Label the 12 undirected Hamiltonian cycles of the complete graph with 5 vertices, 0, 1, ..., 11. Construct the 12×12 (symmetric) matrix D, $D_{i,j}$ giving the distance from cycle i to cycle j defined on the object-space of the *symmetric* TSP. Note that $D_{i,j} \in \{2,3,4,5\}$.

4.4 Branch Merging

A means of abridging a tree scan, in some ways complementary to branch pruning by impasse detection, is *branch merging*. Many search trees possess "equivalent" branches, branches which for search have the same effect or for enumeration have identical weights, hence which may be "merged" and scanned only once. The situation is quite analogous to case analysis theorem proving in which recurring branches of the proof tree are often best established independently as lemmas in order to shorten the analysis. Generally speaking, branch merging is also one of the fundamental means by which the combinatorial analyst derives effective counting formulae. For our purposes, the concept of branch merging is a potent tool often enabling the scientist to eliminate much extraneous computation from his search algorithms.

This section discusses the mechanics, notation, and application of branch merging and its relation to normal tree scan (actually, many algorithms based on this concept do not involve backtracking as such at all). The matter of the dynamic detection of equivalent branches within complex trees is pursued in §§ 6.6 and 7.1.

4.4.1. *The Cut-off and Tier Scans*

We are primarily concerned with branch merging in which equivalent branches initiate from the same level of the tree. Fortunately a large number of problems are of this type. Such merging may conveniently be handled by a *cut-off scan*, a normal tree scan which is interrupted after each j level of analysis to compare partial configurations, choosing only inequivalent configurations from which to continue the analysis. Usually j is a predetermined constant, often equal to 1, but for a sophisticated search might vary from cut-off to cut-off. For $j = +\infty$ a cut-off scan reduces to a normal scan. A cut-off scan may be visualized (for $j = 2$) as in Fig. 4.21. Here, letters are used to label configuration types.

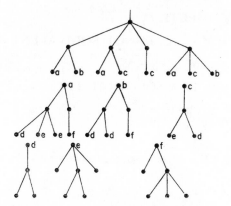

FIG. 4.21. Tree diagram for cut-off scan.

A program structure for a cut-off scan is illustrated in Fig. 4.22 (compare Fig. 4.20). Here, C_k represents some combinatorial object computed at a tree node labeled k, N gives the (old) set of nodes serving as origins for the normal scans through j more levels, and N^* gives the inequivalent terminal nodes of these scans—they become origins for the next cut-off step.

For the case $j = 1$, the nest reduces to a single iteration and the program contains no backtracking at all. This special cut-off scan, which we call the *tier scan*, occurs frequently in generation and enumeration programs in which the equivalence of partial configurations, hence the branches which would spring from them, is detected by a direct comparison (see §§ 4.4.4, 6.6.2, and 7.1). A program structure for this important case appears in Fig. 4.23. The parameter I gives the depth of the tree and the entry N gives the number of inequivalent nodes at the level indexed by i. The procedure *configuration*( ) is a generation procedure which forms in turn each partial configuration C_{i, N_i} (not necessarily new) from the partial configuration $C_{i-1, n}$ formed at the previous level. If the test at line E.6 is successful, C_{i, N_i} already exists in our list of new inequivalent partial configurations. (Actually, since there is no backtracking, only configurations at

$N = \{\text{INITIAL NODE}\}$

\#1 $N^* = \{ \ \}$

for each NODE IN N

    with $i = 1$ to j

    $\cdot \ \cdot \ \cdot$

        for each LINK

           PROGRAM

           $\cdot \ \cdot \ \cdot$

               "We have arrived at a node, call it k"

               COMPUTE C_k

               if $\forall_{h \in N} (C_h$ IS NOT EQUIVALENT TO $C_k)$:

                   $N^* \cup \{k\} \rightarrow N^*$

               if C_k IS A SOLUTION: EXIT

if $N^* \neq \{ \ \}$: $N = N^*$; go to \#1

NO SOLUTION

FIG. 4.22. Generic program structure for cut-off scan.

E.1 $N_0 = 1$

E.2   for $i = 1$ to I:

E.3 $N_i = 0$

E.4   for $n = 0$ to $N_{i-1} - 1$:

E.5   for $configuration(C_{i-1,n}) \rightarrow C_{i,N_i}$:

E.6   if $\exists_{n^*=0 \text{ to } N_i - 1}(C_{i,N_i}$ IS EQUIVALENT TO $C_{i,n^*})$:

E.7   reiterate

E.8   otherwise:

E.9 $N_i + 1 \rightarrow N_i$

FIG. 4.23. Program structure for tier scan.

two levels of the tree, new level and old level, need be preserved. Thus, in practice, a name interchange technique such as described in § 2.1.3 can be employed to save storage —see Fig. 4.24, for instance.)

Apart from branch merging by direct comparison, the cut-off and tier scans are also useful when there is a good chance of successfully completing a search at a shallow level of the tree and we do not wish to spend a lot of time exhausting possibilities at deep levels. The approximation algorithm of § 4.3.4 is essentially of this type. However, even with equivalent branches merged, search trees tend to be broader than they are deep, hence the cut-off scan usually requires considerably more storage than the normal scan.

4.4.2. *Permanent Evaluation*

The merging of equivalent branches of a uniform tree may sometimes be accomplished by analysis prior to computation. This situation is illustrated by permanent evaluation. Let $M_{0 \text{ to } n-1, \, 0 \text{ to } n-1}$ be an $n \times n$ matrix of real numbers. The *permanent* of M may be defined as

$$\sum_{\text{each } permutation(n) \to p} \left(\prod_{j=0 \text{ to } n-1} (M_{j, \, p_j}) \right) \tag{4.3}$$

where *permutation(n)*, hence $p_{0 \text{ to } n-1}$, assumes for its values each of the $n!$ permutations of $0, 1, \ldots, n-1$. Use of (4.3) to compute the permanent of a given matrix is extremely inefficient, however, even when the product evaluation is merged with the permutation generation in order to reduce the number of required multiplications to something less than $n \cdot n!$ (about $en!$). Its fault lies in the repeated scan of many equivalent branches of the search tree: permanents for each of the $\binom{n}{k}$ sub-matrices formed by intersecting k columns with the last k rows are computed $(n-k)!$ times, once for each arrangement of the particular $n-k$ initial columns. This superfluous computation may be omitted by proceeding through the tree level by level, saving the $\binom{n}{k}$ permanents at level k for use in evaluating permanents at level $k+1$ [116].

An algorithm of this sort is given in Fig. 4.24. The structure of this algorithm resembles that of Fig. 4.23, of course. Here, however, the preanalysis obviates an equivalence test; the inequivalent configurations (the combinations) are generated at line P.3 by the generation procedure *combination( )*—see § 5.1.2—rather than by a three-step process as at lines E.4 through E.7 of Fig. 4.23. The quantity $P_{j,k}$ (actually $P_{j(\text{mod } 2),k}$ for efficiency of storage) gives the permanent of the sub-matrix formed by rows $0, 1, \ldots, j-1$ and

P.0   *permanent(n,M)*

P.1   Old $= 0$; New $= 1$; $P_{0,0} = 1$

P.2   for $j = 1$ to n:

P.3   for $k = 0, 1, \ldots,$ as *combination(n,j)* $\to c$:

P.4 $P_{\text{New},k} = \sum_{z \in c} \left(M_{j-1,z} P_{\text{Old, } combtonum(j-1, \, c \sim \{z\})} \right)$

P.5   New $\leftrightarrow$ Old

P.6   *permanent( )* $= P_{\text{Old},0}$

FIG. 4.24. A program for permanent evaluation.

columns specified by elements in the combination of n things taken j at a time whose "serial number"—see § 5.1.3—is k. (Note that we are now building up the sub-matrices beginning at the top of the given matrix.) This permanent is calculated, at line P.4, from the permanents at the previous level corresponding to sub-matrices with one less column; the indices for these permanents are calculated by the procedure *combtonum*( ) discussed in § 5.1.3. A symbolic example of this recursive calculation appears in Fig. 4.25.

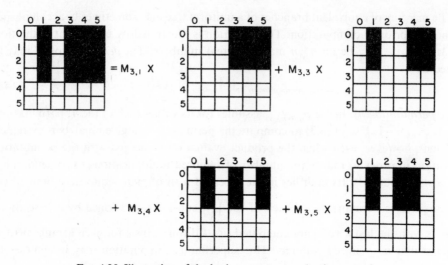

FIG. 4.25. Illustration of the basic permanent evaluation recursion.

4.4.3. *Dynamic Programming Approach to Optimization*

Permanent evaluation as just described is closely related to the so-called "dynamic programming" approach to certain optimization problems. If, in Fig. 4.24, the summation $\left(\sum\right)$ of line P.4 were replaced by a minimization (MIN), then the algorithm would essentially solve a multiplicative assignment problem (see exercise 5 of § 4.3), the number $P_{n(\bmod\,2),0}$ would give the smallest product,

$$\prod_{j=0 \text{ to } n-1} (M_{j,p_j}),$$

over all permutations $(p_0, p_1, \ldots, p_{n-1})$ of 0, 1, $\ldots$, $n-1$. In such a branch merged approach, optimization impasse detection has become the determination of an optimum (in this case, minimum) value from among the various ways in which a partial appraisal (in this case, product) could be formed. Pruned branches of the complete tree correspond to choice of a nonminimum value, a value which need not be considered since a better equivalent partial solution is already known.

For an example, we apply this approach to the TSP (this was first done by Bellman [103] and Held and Karp [113]). First consider the full search tree for the TSP associated with the nest of Fig. 4.15. Such a tree for $n = 6$ is partly drawn in Fig. 4.26. A number attached to a node represents the city visited at that point of the analysis. The expense of traveling from city i to city j, $E_{i,j}$, is associated with a link of the tree; the total cost of a route is the sum of the individual expenses. Branches with the same ancestor set are identical so we need only be concerned with the *best* path from city 0 to city i via cities $k_1, k_2, \ldots, k_j$ over all permutations of these intermediate cities. For instance, in Fig. 4.26 the circled branches are identical—the route to city 2 via cities 1 and 4 cannot affect

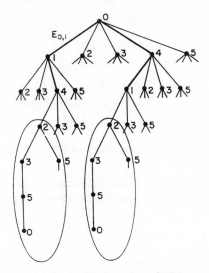

FIG. 4.26. Equivalent branches of TSP tree.

the possible routes through the remaining cities and back to the home city. Thus only the shortest route to city 2 via cities 1 and 4 need be considered.

A dynamic programming algorithm for the TSP is given in Fig. 4.27. Segment T, which has the familiar tier scan structure, recursively computes the length (i.e. cost) of the shortest path from city 0 to city i via the $|c|$ intermediate cities $(c)_1, (c)_2, \ldots, (c)_j$; these lengths are called $P_{i,nbr(c)}$, $i \in n \sim \{0\}$ and $c \subset n \sim \{0,i\}$. The recursion is on j, the number of intermediate cities; the recursive formula is at line T.5 of the program. From these lengths and L, the length of the shortest complete cycle, segment B reconstructs, in inverse order, the sequence of cities, q_h, for $h = 0$ to $n-1$, which form an optimal cyclic tour. (The function $subset(n \sim \{0,i\},j)$ is a generation procedure producing in turn all j-subsets of $\{1,2,\ldots,i-1,i+1,\ldots,n-1\}$; it is similar to $combination(n,j)$ of Fig. 4.24 and is discussed in § 5.1 and presented in § 5.3.1.)

The chief disadvantage of the dynamic programming approach (as with most cut-off scans) is the large amount of storage required. In the TSP example, the 2-array P has $(n-1)(2^{n-1}-1)$ entries. On the other hand, the branch merging tends to shorten the tree scan: only about $n^2 2^{n-2}$ "basic steps" are required in the TSP algorithm. Also, the method is "data independent": the execution time for the algorithm depends only on n and not

T.1   for $z = 1$ to $n-1$: $P_{z,0} = E_{0,z}$

T.2   for $j = 1$ to $n-2$:

T.3   for $i = 1$ to $n-1$:

T.4   for $subset(n \sim \{0,i\}, j) \rightarrow c$:

T.5 $P_{i,nbr(c)} = MIN_{z \in c}(P_{z,nbr(c \sim \{z\})} + E_{z,i})$

T.6 $L = MIN_{z=1 \text{ to } n-1}(P_{z,nbr(n \sim \{0,z\})} + E_{z,0})$

B.1 $q_0, q_n = 0$; $c = n \sim \{0\}$!

B.2   for $h = n-1, n-2, \ldots, 1$:

B.3 $q_h = \text{first } z \in c \ni L = P_{z,nbr(c \sim \{z\})} + E_{z,q_{h+1}}$

B.4 $c \sim \{q_h\} \rightarrow c$; $L - E_{q_h, q_{h+1}} \rightarrow L$

B.5   "$0, q_1, \ldots, q_{n-1}, 0$ is the cycle of cities for"

B.6   "the shortest route"

FIG. 4.27. Dynamic programming algorithm for TSP.

on the entries of the network incidence matrix. Thus though not generally useful in large problems, the dynamic programming approach is often worth considering in moderately sized network problems.

4.4.4. *Enumeration by Branch Merging*

Most combinatorial configurations possess a recursive description which imparts an arborescent structure to their generation—the generation of "Latin squares" (see exercise 11 of § 5.1) by successively adjoining rows to Latin rectangles, for instance. This tree may be extremely irregular, in which case enumeration of the configurations requires their actual generation—that is, enumeration consists of counting terminal nodes in the tree. At the other extreme, the tree may be extremely regular—as with permutations, for instance—and analysis of this regularity (essentially, determination of equivalent branches) yields an enumeration formula. (Such regularity detection is, in fact, what the discipline of enumerative combinatorial analysis is all about!)

By using the tier scan, a detection scheme for equivalent partially constructed configurations through direct comparison, and simple bookkeeping, it is often possible to effect an enumeration which is neither feasible exhaustively nor amenable to combinatorial analysis. The bookkeeping is illustrated in Fig. 4.28, where it is shown how the number of terminal nodes, 22, for the hypothetical complete tree shown in (a) is formed using the tier scan shown in (a). Letters are used to label the nodes. In the tier scan diagram, nodes chosen to represent the origin for a class of equivalent branches are repeated at the end of a dotted line. The numbers written beside these nodes give the number of equivalent ways this partially constructed configuration could have been formed; at the first level this is merely the number of nodes equivalent to the representative node. Such a number is formed by summing the numbers attached to the fathers of equivalent nodes; for instance, the number attached to node f is 3, since node a (father of f) represents 2 branches and node b (father of h, which is equivalent to f) represents 1 branch. The equivalences are given to the right of the trees. Of course, definition and detection of

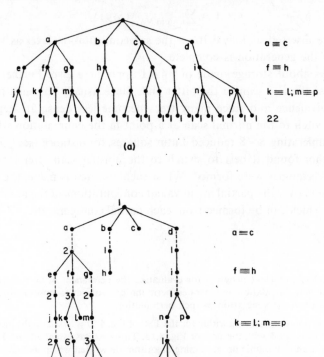

(a)

(b)

FIG. 4.28.

equivalences depends on the particular problem. It should be mentioned that while two equivalent branches necessarily have the same structure, two branches with the same structure are not necessarily equivalent under the particular equivalence relation. In the concocted illustration we did not make node n equivalent to nodes m and p.

A program to accomplish such an enumeration has the form of a tier scan (Fig. 4.23) supplemented by the appropriate bookkeeping. Let $S_{i,n}$ be the number attached to the nth inequivalent node at a level i of the tree. Then, to perform the enumeration, line E.1 of Fig. 4.23 would contain the statement

$$S_{0,0} = 1,$$

line E.9 the statement

$$S_{i,N_i} = S_{i-1,n}$$

(before N_i is incremented) and line E.7 the statement

$$S_{i,n*} + S_{i-1,n} \rightarrow S_{i,n*}$$

(before "reiterate", of course). The final answer would be given by

$$A = \sum_{n=0 \text{ to } N_I - 1} (S_{I,n})$$

which could be inserted as line E.10 (at the same indentation level as line E.2), to be executed when the generation is complete.

The remarks about storage made in § 4.4.1 apply here also. Furthermore, it is not necessary to use the tier scan to the bitter end. Since usually so much time is spent in testing for equivalence and since many trees become less dense at deeper levels, it is often best to switch to the normal scan at a point in the enumeration when it becomes feasible. In enumerating 8×8 reduced Latin squares, for instance (see § 7.1.3 and Wells [254]), the author found it best to switch to the normal scan after inequivalent 5×8 reduced Latin rectangles were formed. When such a switch is made, the final answer is formed as the sum over the partial inequivalent configurations of the number of complete configurations which can be formed from each partial configuration.

EXERCISES

1. Write a timewise efficient program for evaluating the permanent of an incidence matrix (i.e. a zero–one matrix). This calculation enumerates permutations subject to positional restrictions on the marks; § 5.2.1 discusses the generation of such permutations.

2. The dynamic programming algorithm for the TSP of Fig. 4.27 is based on the merging of branches of the search tree associated with the nest of Fig. 4.15. This nest has essentially no Hamiltonian cycle impasse detection. Can a dynamic programming algorithm be written which incorporates Hamiltonian cycle impasse detection in order to shorten either running time or storage requirements of the algorithm?

*3. Using the approach of § 4.4.4, write a program to calculate the number of 6×6 zero–one matrices which have determinant magnitude equal to j, $j = 0$ to 9. Assume the existence of a procedure which determines if two $r \times 6$ 2-arrays of zeros and ones are equivalent under arbitrary row and column permutations—see § 7.1.2.

4. What terminal nodes of Fig. 4.28(a) are counted by the tally 6 at the third terminal node of Fig. 4.28(b)?

CHAPTER 5

GENERATION OF ELEMENTARY CONFIGURATIONS

THIS chapter presents practical schemes for generating the basic elementary combinatorial configurations. Most of these algorithms are given explicitly as generation procedures—procedures designed to produce one configuration after another until all admissible configurations have been formed. However, it is frequently not the generated configurations themselves, but the structure of the generating program which is of primary importance, since (as indicated in Chapter 4, especially in § 4.3) programs to generate points of an object-space form the basis of many combinatorial algorithms. Thus, various procedures for accomplishing the same generation are useful, one scheme rather than another being more adaptable to a particular application. Nevertheless, the controlling factor in the practicality of most generation programs is still the number of configurations being generated. For example, if $n = 14$ and there are no simplifying restrictions, the choice between permutation generation schemes is purely academic (since $14! > 8 \times 10^{10}$).

5.1. Subsets and Combinations

Since the elements of an arbitrary n-set may be labeled with the natural numbers $0, 1, \ldots, n-1$, we may usually restrict ourselves to the generation of subsets of $\{0,1, \ldots, n-1\}$. The class of subsets desired and the order in which they are to be generated determine the structure of the generating scheme to be used (and vice versa). One may need in turn all subsets or just all r-subsets. One may be content with a lexicographic generation or one may require less juggling of elements as successive subsets are formed.

5.1.1. *Generation of All Subsets*

Our language assumes the existence of a library procedure, say *subset(s,S)*, to generate index values specified by the quantifier

$$\text{for } s \subset S: \qquad (5.1)$$

This procedure might be as given in Fig. 5.1. (The resolution of (5.1), including the setting and testing respectively of the *Entry* and *Exit* toggles, is similar to the resolution

S.0 $subset(s,S)$

S.1   if $Entry = 1$: $subset(\ ) = \{\ \}$; go to #1   "first subset"

S.2   if $s = S$: $Exit = 1$; exit from procedure   "final exit"

S.3 $a = (S \sim s)_1$

S.4   . $subset(\ ) = (s \cup \{a\}) \sim a$ "next subset"

S.5   #1 $Exit = 0$; exit from procedure   "intermediate exit"

FIG. 5.1. A subset generation procedure.

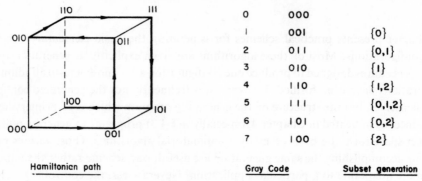

Hamiltonian path	Gray Code	Subset generation
	0 000	
	1 001	{0}
	2 011	{0,1}
	3 010	{1}
	4 110	{1,2}
	5 111	{0,1,2}
	6 101	{0,2}
	7 100	{2}

FIG. 5.2. Illustration of Gray code 3-subset generation.

of "for $i \in I$:" given in Fig. 1.7 with line I.4 now containing the reference "$subset(s,S) \rightarrow s$".) With s viewed as a binary number whose bit positions are determined by the 1-bits in the bit-pattern S, this algorithm essentially adds one to s at each step. The subsets are thus generated in increasing order of $nbr(s)$.

If $S = \{0,1,\ldots,n-1\}$, a simpler generation is possible. The quantifier

$$\text{for } m = 0, 1, \ldots, 2^n - 1: \tag{5.2}$$

yields the subsets of S by means of the transformation

$$s = set(m).$$

By altering the quantifier (5.2), subsets of S may be generated in a different order; for example, use of

$$\text{for } k = 0 \text{ to } log_2(n); \ m \in 2^n \ni |set(m)| = k:$$

would produce (apparently rather inefficiently) subsets in increasing order of cardinality.

Occasionally, one wishes to generate all subsets of n with *minimum* change between successively generated subsets. This corresponds to the tracing of a Hamiltonian path on the n-cube (or, alternatively, to the listing of an n-bit Gray code). An example, for $n = 3$, is given in Fig. 5.2.

$$\text{for } i = 0, 1, \ldots, 2^n - 1:$$
$$s = set(i)$$
$$g = s \triangle (s \ominus \{1\})$$

PROGRAM

FIG. 5.3. A procedure for Gray code subset generation.

$combination(n,r, "C")$

    (real) n

    if $Entry = 1$: $Exit = 0$; $C = (set)\ r$; exit from procedure

    if $r = 0$: $Exit = 1$; exit from procedure

    $A = (n \sim C) \sim (C)_1$

    if $A = \{\ \}$: $Exit = 1$; exit from procedure

    $a = (A)_1$

    $B = (C \sim a) \cup \{a\}$

    $C = B \cup (r - |B|)$; $Exit = 0$; exit from procedure

FIG. 5.4. A combination generation procedure.

A simple algorithm for converting between normal binary representation and Gray code representation appears in § 3.1.2. That conversion can be applied directly as shown in Fig. 5.3; each subset g differs by a single element from its predecessor. This generation does not explicitly display the element being adjoined or deleted. Rather than compute this information from the old and new subsets, it is probably best to extract it directly from s. As can be seen from the proof of (3.4), the element in question is $(s)_1$ and it is being deleted or adjoined according as $(s)_1 + 1$ is in s or not respectively.

For $n > 3$, there exist additional, structurally different n-cube Hamiltonian paths on which a subset generation scheme could be based (see § 4.2.4). Few of them allow the simplicity of the generation scheme of Fig. 5.3, however, nor do they possess the pleasant property that each dimension is completely traversed before a new dimension is entered.

5.1.2. *r-Subsets—Combinations of Distinct Objects*

More commonly one needs to generate not all subsets of $\{0,1,\ldots,n-1\}$ but only those containing r elements, $0 \leqslant r \leqslant n$. This generation is only slightly more involved than the one for all subsets of an arbitrary set (Fig. 5.1); it appears in Fig. 5.4. Given an r-combination C, this procedure calculates the next "largest" r-combination (the ranking is by set serial number—see Fig. 1.3) as the new value of C. Thus a quantifier such as

$$\text{for each } combination(M,q, "C"):$$

could be used to produce in turn each q-combination C of M objects.

$\text{old } C = 0100110$

$(C)_1 = 1$

$A = 1011001 \sim 0000001 = 1011000$

$a = 3$

$B = (0100110 \sim 0000111) \cup \{3\} = 0101000$

$r - |B| = 1$

$\text{new } C = 0101000 \cup (0000001) = 0101001$

Fig. 5.5. Program trace for combination generation.

To understand this procedure it is perhaps best to view the sets in their bit-pattern representation. A sample calculation trace for $n = 7$ and $r = 3$ is given in Fig. 5.5. In words this process may be described as follows: Given a combination C, the next combination is formed from C by moving the rightmost 1-bit which has a 0-bit on its left one place to the left, and then placing all 1-bits to the right of this as far right as possible. The generation is complete when all 1-bits are on the far left. Note that when the jth 1-bit first moves to position k, all combinations with the j rightmost 1-bits in positions

$0, 1, \ldots, k-1$, $\left. \binom{k}{j} \right/$ in number, have already been formed.

This scheme produces combinations in increasing order of $nbr(C)$. Note that this is not the same as increasing lexicographic order of the vector $((C)_1, (C)_2, \ldots, (C)_r)$. The relationship between these orderings is pursued in the exercises (as well as by the next program).

Two additional generation procedures are given in Fig. 5.6. The procedure of segment L is strictly numerical, generating combinations in the list representation; that is, a combination is given as a vector $(c_0, c_1, \ldots, c_{r-1})$, where $c_i \in n$ and $c_i < c_{i+1}$. This algorithm generates r-combinations, considered as r-subsets, in increasing lexicographic order of the dual reverse combination (see § 3.3.1). The procedure of segment N also generates combinations in the list representation. The recursive structure is the important feature of this program. (The dummy statement "$i = -1$" at line N.8 is primarily inserted into the program to accommodate the transfer of control necessary when $Entry = 0$. The necessity of such statements in generation procedures apparently indicates a deficiency in the language with respect to the nest notation.)

5.1.3. Serial Numbers and Ranking of Combinations

As mentioned in § 3.2.1, it is often convenient to represent each member of a class of configurations by a distinct easily calculable natural number index. For most of the configurations discussed in this chapter, this is accomplished by a natural ranking of the configurations. The position in the ranking (beginning with 0) is the assigned index, the *serial number* for the associated configuration. For example, subsets of $\{0, 1, \ldots, n-1\}$ are naturally ranked as indicated in Fig. 1.3, with $nbr(s)$ yielding the serial number for the subset s.

L.0 $combination(n,r,``c")$
L.1   if $Entry = 1$:
L.2   for $j \in r : c_j = j$
L.3 $Exit = 0$; exit from procedure
L.4   for $j = r-1, r-2, \ldots, 0$:
L.5   if $c_j < n-r+j$:
L.6 $c_j + 1 \rightarrow c_j$
L.7   for $k = j+1$ to $r-1$:
L.8 $c_k = c_{k-1} + 1$
L.9 $Exit = 0$; exit from procedure
L.10 $Exit = 1$; exit from procedure

N.0 $combination(n,r,``c")$
N.1   if $Entry = 0$: go to $\#\,1$
N.2 $c_r = n$
N.3   with $i = r-1, r-2, \ldots, 0$;
N.4 $\ddots$
N.5   for $c_i = i$ to $c_{i+1} - 1$:
N.6 $\ddots$
N.7 $Exit = 0$; exit from procedure
N.8 $\#\,1$ $i = -1$ "continue"
N.9 $Exit = 1$; exit from procedure

Fig. 5.6. Two signature combination generation procedures.

The r-combinations of $\{0,1,\ldots,n-1\}$ may also be ranked naturally, in the order generated by the procedure of Fig. 5.4. The serial number N is a simple sum, shown in Fig. 5.7 as the result of the $combtonum(\ )$ procedure. (While this calculation consists of the evaluation of a single formula, it is given here as a procedure since it is often called from many places in a program. To avoid time-consuming transmission of arguments for this simple calculation, we assume that it appears in a program as a brother to segments containing its references.) The truth of this formula is seen by counting the number of bit-pattern combinations appearing before C in the generation sequence. (Also, see exercise 5.) Note that although $0 \leqslant N < \binom{n}{r}$, n is not needed as an argument of the procedure.

The inversion of this formula, i.e. the computation of the combination C associated with the serial number N, appears as the $numtocomb(N,m,n)$ procedure in Fig. 5.7. This procedure has the same basic structure as the conversion given in Fig. 3.1, but in this case a simple division will not produce the largest value of k for which $\binom{k}{j} \leqslant N$—that value must be found by trial (line C.3).

N.0   *combtonum*( ) "combination to serial no. conversion"

N.1 $combtonum( ) = \sum_{j=1 \text{ to } r} \left[\binom{(C)_j}{j} \right]$

N.2   exit from procedure

C.0   *numtocomb*(N,n,r) "serial no. to combination conversion"

C.1 $C = \{\ \}; k = n-1$

C.2   for $j = r, r-1, \ldots, 1$:

C.3   (first $h = k, k-1, \ldots \ni N - \binom{h}{j} \geq 0) \rightarrow k$

C.4 $C \cup \{k\} \rightarrow C; N - \binom{k}{j} \rightarrow N$

C.5   *numtocomb*( ) = C; exit from procedure

FIG. 5.7. Conversions between combinations and their serial numbers.

EXERCISES

1. Describe the configurations s generated by the procedure of Fig. 5.8.

Q.0   *thing*(n,"s")

Q.1   if *Entry* = 1: $s = \{\ \}$; *Exit* = 0; exit from procedure

Q.2   for $i = 0$ to $n-1$:

Q.3   if $i \notin s$: $s \cup \{i\} \rightarrow s$; *Exit* = 0; exit from procedure

Q.4 $s \sim \{i\} \rightarrow s$

Q.5   *Exit* = 1

FIG. 5.8. What is being generated?

2. Modify the program in Fig. 5.4 so that it produces combinations in the same order as produced by segment L of Fig. 5.6. (Hint: What is the effect on the bit-patterns in the Fig. 5.4 program of replacing r by $n-r$?)

3. (a) Modify segment L of Fig. 5.6 to produce combinations in decreasing order of $nbr(\{c_0,c_1,\ldots, c_{r-1}\})$.

(b) Describe the order of combinations produced by segment N of Fig. 5.6.

4. As indicated in § 3.3.1, an r-combination of $\{0,1,\ldots,n-1\}$ with object repetition allowed may be represented by an $n+r-1$ bit bit-pattern containing r 1-bits.

(a) Write a program to produce a vector $(c_0,c_1,\ldots,c_{r-1})$ from a set C representing such a combination.

(b) Write a strictly numerical combination-with-repetition generation procedure.

5. Express the serial number formula for an r-combination of n (line N.1 of Fig. 5.7) in terms of the list representation

$$(c_0,c_1,\ldots,c_{r-1}), c_{i-1} < c_i, \text{ for } i = 1 \text{ to } r-1.$$

(This very natural indexing scheme has been discovered and used by several combinatorial researchers—see Lehmer [143].)

6. Devise a formula for the serial number of an r-combination of n with repetition allowed when the combinations are ordered as sets, e.g. when $\ldots < (0,1,1) \equiv 01101 < (1,1,1) \equiv 01110 < (0,0,2) \equiv 10011 < \ldots$.

*7. It is possible to generate r-combinations of n so that successive combinations differ as little as possible, by exactly two elements. The 35 3-subjects of $\{0,1,\ldots,6\}$ are so listed in Fig. 5.9, for example. Find the key to this generation; that is, write a procedure to accomplish such generations in general.

1.	0000111	13.	1010001	25.	0001110
2.	0001011	14.	0110001	26.	0011100
3.	0010011	15.	1100001	27.	0101100
4.	0100011	16.	1100010	28.	1001100
5.	1000011	17.	1010010	29.	1010100
6.	1000101	18.	1001010	30.	0110100
7.	0100101	19.	1000110	31.	1100100
8.	0010101	20.	0100110	32.	1101000
9.	0001101	21.	0101010	33.	1011000
10.	0011001	22.	0110010	34.	0111000
11.	0101001	23.	0011010	35.	1110000
12.	1001001	24.	0010110		

FIG. 5.9. Example of "minimum-change" r-combination generation.

8. (a) Write a procedure to generate r-subsets of an arbitrary given set S.

(b) Write a procedure for finding the serial number of such a subset assuming a lexicographic indexing scheme.

9. Let $a_0, a_1, \ldots, a_m$ be a given sequence of positive integers. For $n = m+1$ to N, let

$$a_n = MIN_{s \subset n-1 \ni s \neq \{\ \}} \left(\prod_{j \in s} (a_j) + \prod_{j \in n \sim s} (a_j) \right).$$

Write an efficient program to form $a_{m+1}, a_{m+2}, \ldots, a_N$.

10. Write a program to form a random r-combination of n. (Note: When we speak of a "random" configuration we mean that the configuration could with equal probability be any one of the configurations in question. For example, a random element of $\{0,1,\ldots,99\}$ could be formed by the statement "$r = [100 \ random]$"—see § 2.2.4.)

*11. A *Latin square of order n* is an n by n 2-array of the labels $0, 1, \ldots, n-1$ such that no label appears more than once in each row or in each column. Two Latin squares A and B are *orthogonal* if $(A_{h,i}, B_{h,i}) = (A_{j,k}, B_{j,k})$ only when $h = j$ and $i = k$. A theorem of Mann (see [146]) says that for a Latin square of order $4n+2$ to have an orthogonal mate, it is necessary that every $2n+1$ by $2n+1$ subsquare contain every set of $2n+1$ labels at least $n+1$ times. Write a program which applies this criterion to a given Latin square.

12. Segment S of Fig. 5.10 generates all subsets S of $\{0,1,\ldots,I-1\}$ by a recursive technique.

(a) Describe the order in which subsets are generated.

(b) Modify this program to include arbitrary given restrictions (i.e. impasse detections) of the form "element j cannot coexist in a subset with element k".

13. Describe the sets $\{c_1,c_2,\ldots,c_r\}$ produced by segment K of Fig. 5.10, where $c_0 = 0$.

14. Write an efficient procedure which produces subsets of $\{0,1,\ldots,n-1\}$ in non-decreasing order of cardinality.

S.1   with $i = 0$ to $I-1$, as $S = \{\ \}, \{\ \} \cup q_{i-1}, \ldots$:

S.2 $\cdot\cdot$

S.3   for $q_i \subset \{i\}$:

S.4 $\cdot\cdot$

S.5   USE S

K.1   with $i = 1$ to r, as $T = K, K - c_{i-1}, \ldots$:

K.2 $\cdot\cdot$

K.3 $A_i = T - (N-r+1)(r-i) - \begin{pmatrix} r-i+1 \\ 2 \end{pmatrix}$

K.4 $B_i = \dfrac{T - \begin{pmatrix} r-i+1 \\ 2 \end{pmatrix}}{r-i+1}$

K.5   for $c_i = max(c_{i-1}+1, A_i)$ to B_i:

K.6 $\cdot\cdot$

K.7   USE $\{c_1, c_2, \ldots, c_r\}$

FIG. 5.10. Recursive subset generation.

15. Suppose we have F configurations each of which may possess certain properties $p_0, p_1, \ldots, p_{n-1}$. If for each S, $S \subset n$, we know $f(S)$, the number of configurations which possess properties p_j, $j \in S$, then we may calculate the number of configurations N which possess none of the properties by the principle of inclusion and exclusion [36], which says

$$N = \sum_{k=0 \text{ to } n-1} \left[(-1)^k \sum_{S \subset n \ni |S|=k} (f(S)) \right],$$

where $f(\{\ \}) = F$. Assuming the existence of the function f, write an efficient program for the evaluation of this formula.

16. Let $D \subset C$. What does the formula

$$\sum_{i=1 \text{ to } |C| \ni (C)_i \in D} (2^{i-1})$$

express?

5.2. Permutations of Distinct Objects

We consider here permutations $(p_0, p_1, \ldots, p_{n-1})$ of the n distinct marks (i.e. labels) $0, 1, \ldots, n-1$. These marks may, of course, be used to index an arbitrary n-set of objects. The more general case in which the marks are not necessarily distinct is treated in § 5.3.

Of the many permutation generating schemes appearing in the literature, we present only two types: backtrack schemes which generate the permutations in lexicographic order and schemes in which successively generated permutations differ by a single transpositions of marks. Procedures of the first type are important because of their straightforwardness and general applicability. Transposition procedures are important since they require minimum juggling of information for successive arrangements; they are useful in problems involving method-of-ascent (see § 4.4.3), for instance. Additional permutation generation schemes are discussed in Lehmer [64] as well as in reference [11].

5.2.1. *Lexicographic Generation and Serial Numbers*

A generation procedure which produces permutations in increasing lexicographic order of the signatures is given in Fig. 5.11. Here, for the sake of efficiency, the upper limit on the next index is $n-2$ and p_{n-1} is calculated independently; in general, it is

P.0	*permutation*$(n, \text{"}p\text{"})$	"$n \geqslant 1$"
P.1	if *Entry* $= 0$: go to #1	
P.2	$S = (\text{set}) \, n$	
P.3	with $i = 0$ to $n-2$:	
P.4		
P.5	for $p_i \in S$:	
P.6	$S \sim \{p_i\} \to S$	
P.7		
P.8	$p_{n-1} = (S)_1$	
P.9	*Exit* $= 0$; exit from procedure	
P.10	#1	$S = \{p_{n-1}\}; \, i = n-1$
P.11	$S \cup \{p_i\} \to S$	
P.12	*Exit* $= 1$; exit from procedure	

FIG. 5.11. Basic recursive permutation generation.

often more convenient to let the nest index control the *entire* generation. (See exercise 6(a) of § 4.1 for a notational variant of another sort.) This procedure should be compared with recursive combination generation (segment N of Fig. 5.6) to note the structural difference of signature generation programs produced by the relevance of the order of the components.

This basic algorithm may readily be modified to generate permutations subject to various restrictions. For instance, for "derangement" [37] generation, the statement

$$\text{for } p_i \in S \text{ such that } p_i \neq i:$$

could replace the quantifier at line P.5. More generally, suppose that the allowable values for p_i are given by a mask R_i; that is, $p_i \in R_i$ must be true for each $i \in n$. (In combinatorics, the complement of the zero–one matrix R is often called the *board* of restrictions.) Then, replacing the quantifier by

$$\text{for } p_i \in S \cap R_i:$$

and line P.8 by

$$p_{n-1} = (R_{n-1} \cap S)_1$$

produces a procedure which generates permutations subject to the positional restrictions established by R. (See exercise 2.)

The search tree for the program of Fig. 5.11 is essentially identical to that given in Fig. 4.3 for the factorial digit generation of Fig. 4.2. An association between these factorial digits $a_0, a_1, \ldots, a_{n-1}$ (see § 3.1.1) and the permutations is given by

$$\text{for } i = 0 \text{ to } n-1: p_i = \left(n \sim \bigcup_{j \in i} [\{p_j\}] \right)_{a_{n-1-i}+1}. \tag{5.3}$$

For example, if $a_0 = 0$, $a_1 = 0$, $a_2 = 2$, $a_3 = 1$, then $p_0 = (\{0,1,2,3\})_2 = 1$, $p_1 = (\{0,1,2,3\} \sim \{1\})_3 = 3$, etc., so that $p = (1,3,0,2)$. The relationship between a set of factorial digits is an index N, $0 \leqslant N < n!$ is

$$N = \sum_{i=1 \text{ to } n-1} (a_i i!). \tag{5.4}$$

Thus we have a natural ranking for the permutations. This serial number calculation is given as the *permtonum*( ) procedure of Fig. 5.12; the inverse calculation is the *numtoperm*( ) procedure. These procedures are direct applications of (5.3) and (5.4) and the conversion technique discussed in § 3.1.1.

N.0   *permtonum*(p)

N.1   *permtonum*( ) =

N.2 $\sum_{i=0 \text{ to } n-2, \text{ as } S=n, \, n\sim\{p_{i-1}\}, \ldots} [\,|\,p_i \cap S|\,(n-1-i)!\,]$

N.3   exit from procedure

P.0   *numtoperm*(N,n)

P.1   for $i = 0$ to $n-1$, as $S = $ (set) $n, \ldots, S \sim \{p_{i-1}\} \rightarrow S, \ldots$:

P.2   define X, N by $\dfrac{N}{(n-1-i)!}$

P.3 $p_i = (S)_{X+1}$

P.4   *numtoperm*( ) = p

FIG. 5.12. Conversions between permutations and their serial numbers.

5.2.2. *Generation by Transposition*

The lexicographic generation procedure of Fig. 5.11 shuffles more than two marks half of the time. Procedures are available which require but a single interchange of two marks to form each permutation from its predecessor. One such algorithm is given in Fig. 5.13. This procedure has the same tree structure as factorial digit generation (Fig. 4.2); in fact, certain properties of the vector of factorial digits $(a_{n-1}, a_{n-2}, \ldots, a_1)$ are used to determine the necessary transposition.

W.0		*permutation*$(n, \text{"}p\text{"}, \text{"}a\text{"})$
W.1		if *Entry* $= 0$: go to $\# 1$
W.2		$a_n = 0$; for $i = 0$ to $n-1$: $p_i = i$
W.3		with $i = n-1, n-2, \ldots, 1$:
W.4		$\cdot \cdot$
W.5		$\quad$ for $a_i = 0, 1, \ldots, i$:
W.6		$\quad\quad$ if $a_i = 0$: nest deeper
W.7		$\quad\quad$ if i is odd or $a_{i+1} < 2$: $h = i-1$
W.8		$\quad\quad$ otherwise: $h = max(0, i-a_{i+1})$
W.9		$\quad\quad p_h \leftrightarrow p_i$
W.10		$\cdot \cdot$
W.11		$\quad\quad\quad$ *Exit* $= 0$; exit from procedure
W.12	$\# 1$	$i = 0$
W.13		$\quad$ *Exit* $= 1$; exit from procedure

FIG. 5.13. Wells method for permutation-by-transposition generation.

Upon initial entry (i.e. when *Entry* $= 1$), the identity permutation $(0, 1, \ldots, n-1)$ and the corresponding nest vector $(0, 0, \ldots, 0)$ are produced. At later entries control immediately passes to the center of the nest where backtracking is initiated in search of the smallest i for which $a_i \neq i$. We reach line W.7 with this value for i (the generation is complete when we have $a_i = i$, for all i). At W.7 all permutations of marks $p_j, j \in i$, given a fixed arrangement $p_i, p_{i+1}, \ldots, p_{n-1}$, have been generated, and we are about to interchange mark p_i with an "earlier" mark. The rules expressed by the test of lines W.7 and W.8 insure that each time this interchange takes place during a subgeneration for marks $(p_0, p_1, \ldots, p_i)$ p_i receives a different value; this guarantees that all permutations will be generated (a detailed proof of this may be found in Wells [152]).

Many modifications of this procedure are possible; we mention two. First, as pointed out by Heap [141], some time-saving is possible by precomputing a table of interchange indices—an interchange schedule. For instance, lines W.7 and W.8 could be replaced by the formula "$h = H_{i,a_{i+1}}$" where the 2-array H is precomputed from the given rules. (It should be mentioned that different rules exist which still guarantee the exhaustive generation of $n!$ permutations as outlined in the previous paragraph.)

A second variation concerns omitting branches of the generation. Suppose that following the interchange of marks p_h and p_i, $h < i$, it is discovered that the next $i!$ permutations—those involving rearrangements of the marks $p_0, p_1, \ldots, p_{i-1}$—are not needed. By application of a transposition when i is odd or a cyclic permutation when i is even, we may immediately skip to that permutation upon which the next interchange involving p_i may be performed (when a_i is increased by one). The transposition to be applied when i is odd ($i > 1$) is

$$(p_{i-1}p_j), \text{ where } j = \begin{cases} i-2, & \text{if } a_i < 2 \\ \max(0, i-1-a_i), & \text{otherwise.} \end{cases} \quad (5.5)$$

The cycle to be applied when i is even is

$$(p_1 p_0) \text{ for } i = 2,$$
$$(p_2 p_1 p_3 p_0) \text{ for } i = 4,$$

and

$$(p_1 p_3 \cdots p_{i-5} p_{i-2} p_2 p_4 \cdots p_{i-4} p_{i-3} p_{i-1} p_0) \text{ for } i \geqslant 6. \quad (5.6)$$

These permutations of $(p_0, p_1, \ldots, p_{i-1})$ are the same as those effected by the actual subgenerations; that, in fact, is the source of their derivation.

5.2.3. *Generation by Adjacent Transposition*

A transposition algorithm of a different sort was discovered independently by Johnson [142] and Trotter [150]. This scheme (which has its tree structure for $n = 4$ illustrated in Fig. 5.14) generates a new permutation by interchange of *adjacent* marks of the preceding permutation.

The idea of the scheme is as follows: Suppose we have the $(n-1)!$ $(n-1)$-permutations of the marks $1, \ldots, n-1$. Each one of these yields n permutations of the marks $0, 1, \ldots, n-1$ by placing mark 0 in each of the n possible positions relative to the permutation $(p_1, p_2, \ldots, p_{n-1})$. In the algorithm, this is accomplished by moving the mark 0 from one end of an $(n-1)$-permutation to the other end in $n-1$ steps, each step consisting of an interchange with a neighboring mark. When mark 0 reaches an end position, the next $(n-1)$-permutation is formed and mark 0 steps back across to the other end, and so forth. The $(n-1)!$ permutations of the marks $1, 2, \ldots, n-1$ are formed in this same way using each of the $(n-2)$ permutations of marks $2, 3, \ldots, n-1$ through which

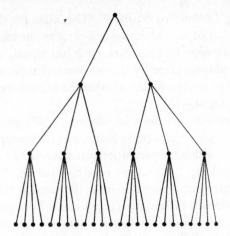

FIG. 5.14. Tree structure of Johnson–Trotter generation.

J.0	$permutation(n, \text{``}p\text{''}, \text{``}t\text{''}, \text{``}d\text{''})$
J.1	if $Entry = 1$:
J.2	$p_0 = 0$
J.3	for $k = 1$ to $n-1$: $t_k = 0$; $d_k = +1$; $p_k = k$
J.4	$Exit = 0$; exit from procedure
J.5	$v = 0$
J.6	for $k = n-1, n-2, \ldots, 1$:
J.7	$t_k + d_k \rightarrow t_k, q$
J.8	if $q = k+1$: $d_k = -1$; reiterate
J.9	if $q = 0$: $d_k = +1$; $v+1 \rightarrow v$; reiterate
J.10	$p_{q+v} \leftrightarrow p_{q+v-1}$
J.11	$Exit = 0$; exit from procedure
J.12	$Exit = 1$; exit from procedure

FIG. 5.15. Johnson–Trotter method for permutation-by-adjacent-transposition generation.

to step mark 1. This process is repeated for the $(n-2)$-permutations, the $(n-3)$-permutations and so forth to the 2-permutations where the interchange of marks $n-1$ and n signals that half of the generation is complete. Relative to the tree, passage from node j_k to node j_k+1 at level k ($j_k = 0$ to $k-1$ for $k = 1$ to $n-1$) corresponds to the interchange involving mark $n-k-1$ which puts it in the (j_k+1)-st position across the current permutation of marks $n-k, n-k+1, \ldots, n-1$.

A Johnson–Trotter algorithm is given in Fig. 5.15 using the generation-without-nest notation discussed in § 4.1.3. The iteration at lines J.6 through J.11 searches for the deepest level containing "unfinished business". The index for this level k supplies the mark $n-k-1$, which is to be moved across one step by the transposition at line J.10. The position of this mark depends upon how far it has already progressed in its movement across the k-permutation, given by t_k, and where that permutation is, given by v. The quantities d_k, $k = 1$ to $n-1$, indicate whether the mark is being moved to the right ($d_k = +1$) or to the left ($d_k = -1$).

There are two important variations to schemes which generate permutations by transposition. First, when the objects being permuted are cumbersome, it is best to omit the explicit references to p within the generation procedures and to treat the transposition indices themselves, (h, i) of Fig. 5.13 and $(q+v, q+v-1)$ of Fig. 5.15, as the primary output. Second, since the parity of a permutation is the same as the parity of the number of transpositions used when the permutation is represented as a product of transpositions (see exercise 3 of § 3.3), transposition schemes alternately generate even and odd permutations. Thus an even (or odd) permutation generation procedure results by remaining within a transposition procedure for two steps.

5.2.4. *Permutations Consistent with a Partial Ordering*

We now impose precedence restrictions upon the marks and construct an algorithm to generate permutations that remain consistent with these restrictions.

Let $A_{0 \text{ to } n-1}$ be a 1-array of subsets of $\{0, 1, \ldots, n-1\}$ which satisfy the following conditions of consistency (asymmetry) and transitivity:

    (1) $i \neq j$ and $i \in A_j \Rightarrow j \notin A_i$.

    (2) $j \in A_i$ and $k \in A_j \Rightarrow k \in A_i$.

Let $B_{0 \text{ to } n-1}$ be the transpose of A, i.e. $i \in A_j \Leftrightarrow j \in B_i$. Such precedence relations yield a *partial ordering* of the marks $0, 1, \ldots, n-1$; that is, the set $\{0, 1, \ldots, n-1\}$ and the relations $A_{0 \text{ to } n-1}$ form a *partially ordered set* (or *poset*—see § 6.3.3).

We can view A_i as the set of marks which must follow mark i (i.e. have a larger signature index) and B_i as the set of marks which must precede mark i in a consistent permutation. For example, if $n = 3$, $A_0 = \{1\}$, and B_0, A_1, B_1, A_2, and $B_2 = \{ \}$, the consistent permutations are $(0,1,2)$, $(0,2,1)$, and $(2,0,1)$. The construction of a transitive system from an arbitrary set of relations and/or the detection of consistency of the relations are pursued in § 6.3. An algorithm for merely counting permutations consistent with a given poset is described in § 7.2.3.

The number of permutations consistent with a partial ordering is usually small compared with $n!$ (a single relation reduces the number of permissible permutations to $n!/2$). It would therefore be grossly inefficient to generate each of the $n!$ permutations and *then* test for consistency—we should let the precedence relations themselves guide the generation. This may be accomplished as in Fig. 5.16.

This procedure has the basic structure of lexicographic generation (Fig. 5.11), only here there is significant impasse detection due to the precedence restrictions. The set S,

P.0 $precedent(n,A,B,"p","R")$

P.1   (set) A, B

P.2   if $Entry = 0$: go to #1

P.3 $S = $ (set) n; for j: $R_j = (n-|A_j|) \sim |B_j|$

P.4   with $i = 0$ to $n-1$:

P.5   . .

P.6 $M_i = \{ \}$; $t = 0$

P.7   for $j \in S \ni i \in R_j$ and $B_j \cap S = \{ \}$:

P.8   if $|R_j \sim i| = 1$:

P.9   if $t = 1$: backtrack

P.10   otherwise: $t = 1$; $M_i = \{j\}$

P.11   if $t = 1$: reiterate

P.12 $M_i \cup \{j\} \to M_i$

P.13   for $p_i \in M_i$:

P.14 $S \sim \{p_i\} \to S$

P.15   . .

P.16 $Exit = 0$; exit from procedure

P.17   #1 $S = \{ \}$

P.18 $S \cup \{p_i\} \to S$

P.19 $Exit = 1$; exit from procedure

FIG. 5.16. Generation of poset consistent permutations.

as before, gives the marks which have yet to be assigned a position in the permutation. However, the use in the ith position of some of these marks is inconsistent with the restrictions, hence a subset of S, M_i, from which the nest vector component p_i can be chosen, is formed at lines P.6 through P.12.

First of all, note that mark j cannot occupy positions $0, 1, \ldots, |B_i|-1$ or positions $n-|A_i|, n-|A_i|+1, \ldots, n-1$, since these areas must be reserved respectively for those marks which precede and follow mark j. The precomputed (at line P.3) set R_j, $j \in n$, gives the positions not so blocked to mark j. Also, a mark j is available for use in position i only if all marks which must precede it already exist in the partial permutation. The quantifier at line P.7 restricts the elements of M_i according to these observations. One further impasse detection is incorporated into this loop which forms M_i. The test at line P.8 asks if this is the last chance for placing mark j; if so, it becomes the sole element in P_i (line P.10). If position i is the last chance for yet another mark (test at line P.9 is successful), an impasse exists.

EXERCISES

1. Write a program to generate solutions to the *problem of the queens*, that is, to generate ways in which n queens may be placed on an $n \times n$ chessboard so that no two queens attack each other.

2. Write a restricted position permutation generation program which detects for the following impasses: (a) there exists an unplaced mark with no empty positions available to it, and (b) there exists an empty position with no unplaced marks to occupy it. (An implementation of this elaboration on the Maniac II computer reduced the total generation time for a 10 by 10 problem with 44 restrictions by a factor of 2. Still better impasse detection can be based on the theorem given in exercise 9 of §6.4.)

3. What is the permutation of $(0,1,2,3,4,5)$ associated with the serial number 389? What is the permutation of $(0,1,\ldots,7)$ associated with this same number?

4. List the 24 permutations of $(0,1,2,3)$ in the order induced by increasing order of the associated factorial digits treated as a vector (a_0,a_1,a_2,a_3)—e.g. $(0,0,0,0) < (0,0,0,1) < \ldots < (0,1,2,3)$.

5. Write down the sequences in which the Wells and Johnson–Trotter schemes generate permutations of four objects.

6. Draw a representation of the permutation object-space (for $n = 3$) as a graph where permutations are adjacent if they differ by a transposition. How many generation-by-transposition schemes exist for $n = 3$?

7. Modify the program of Fig. 5.13 to include the branch-skipping rules (5.5) and (5.6) by introducing another input argument, say g, indicating the acceptance or rejection of the previous permutation. For instance, $g = 0$ might say that the previous permutation was acceptable, generate the next permutation as usual, while $g = 1$ could indicate that the mark which just arrived in position i (as a result of the transposition $(p_h p_i)$, $h < i$) makes that permutation unacceptable, generate the next permutation which has a different mark in position i.

8. Program the Johnson–Trotter algorithm recursively using a nest of iterations.

9. Write a procedure to generate only odd permutations.

10. Write a procedure to generate r-permutations of $\{0,1,\ldots,n-1\}$. (There are at least two distinct useful approaches to this generation.)

11. Write a program to generate permutations selected at random from the permutation object-space. (In this respect see de Balbine [137].)

12. Application of the following rules generates the lexicographically next permutation from a given permutation $(p_0,p_1,\ldots,p_{n-1})$:

(a) Find largest i such that $p_{i-1} < p_i$.

(b) Find largest j such that $p_{i-1} < p_j$.

(c) Interchange p_{i-1} and p_j.

(d) Reverse the order of $p_i, p_{i+1}, \ldots, p_{n-1}$.

Write a generation procedure which applies these rules. (According to Hall and Knuth [22], this method was first published by Fischer and Krause in 1812 [139]. More recently it has been proposed by Shen [148].)

*13. Write an efficient program to calculate the number f_n of permutations $p_{1\ \text{to}\ n}$ of $(0,1,\ldots,n-1)$ which satisfy the following property (where $a \equiv p_i \triangle p_{i+1}$ and $b \equiv p_j \triangle p_{j+1}$):

$$\mathbf{\forall}_{i=1\ \text{to}\ n-3;\ j=i+2,\ i+4,\ \ldots,\ n-1}(a \cap b = 0 \text{ or } a \sim b = 0 \text{ or } b \sim a = 0)$$

is true. This is the number of distinct ways in which a strip of n distinguishable stamps can be folded; for example, $f_1 = 1$, $f_2 = 2$, $f_3 = 6$, and $f_4 = 16$. It is not difficult to show that for $n > 2$, $f_n = 2ng_n$.

A table of g_n, $n = 3$ to 13, is given in Appendix II (Table II.3). (A more extended table appears in Lunnon [145].)

*14. Devise a permutation generation scheme which satisfies the following rules (see Papworth [147]):

(a) All $n!$ permutations are generated and the scheme is cyclic, i.e. application of the algorithm to the last generated permutation produces the first permutation.

(b) In any permutation no mark may occupy a position more than one place removed from its position in the previous permutation.

(c) No mark may remain in the same position for more than two consecutive permutations.

15. List the permutations of $\{0,1,2,3,4,5\}$ which are consistent with the poset $A_0 = \{4\}$, $A_1 = \{5\}$, $A_2 = \{4\}$, $A_3 = \{1,4,5\}$, $A_4 = \{\ \}$, and $A_5 = \{\ \}$. (The *Hasse diagram*

is a convenient pictorial representation of this poset.)

5.3. Permutations with Repeated Objects

Suppose now that the objects being permuted are not distinct. Since we can always make a preliminary scan of the objects we assume a known distribution, namely n objects altogether, m_0 0's, m_1 1's, ..., m_{k-1} $(k-1)$'s; $n = \sum_{i \in k}(m_i)$. As in the previous section, the reader may easily supply the necessary modifications for the case where the objects being permuted are not the natural numbers 0, 1, ..., $k-1$. It is common terminology to say that the n objects have *specification m*, where m is a partition of n.

5.3.1. *Lexicographic Generation*

Given k and the signature $m_{0 \text{ to } k-1}$ (hence n), there are

$$\frac{n!}{m_0!\, m_1! \ldots m_{k-1}!} \tag{5.7}$$

permutations to be generated. This multinomial coefficient may be written, more suggestively, in the form

$$\binom{n}{m_0}\binom{n-m_0}{m_1}\binom{n-m_0-m_1}{m_2} \cdots \binom{m_{k-2}+m_{k-1}}{m_{k-2}}\binom{m_{k-1}}{m_{k-1}}. \tag{5.8}$$

Here we notice that to form a permutation we first choose m_0 out of n empty positions in which to place the m_0 zero marks. We then choose m_1 from the remaining $n-m_0$ empty positions for the ones, m_2 from the then $n-m_0-m_1$ empty positions for the twos, and so forth. The generation of all possible permutations is then a straightforward backtrack program as given in Fig. 5.17. The empty positions are given by S_i, $i = 0$ to $k-1$; the

P.0	$gperm(n,k,\text{“}m\text{”},\text{“}p\text{”})$
P.1	if $Entry = 0$: go to #1
P.2	$S_0 = $ (set) n
P.3	with $i = 0$ to $k-1$:
P.4	$\cdot$ $\cdot$
P.5	for each $rsubset(S_i, m_i, T_i) \rightarrow T_i$:
P.6	for $j \in T_i$: $p_j = i$
P.7	$S_{i+1} = S_i \sim T_i$
P.8	$\cdot$ $\cdot$
P.9	$Exit = 0$; exit from procedure
P.10	#1 $i = k-1$
P.11	$Exit = 1$; exit from procedure
P.R.0	$rsubset(S,r,C)$
P.R.1	if $Entry = 1$:
P.R.2	if $r = 0$: $C = \{\ \}$
P.R.3	otherwise: $C = S \cap [(S)_r + 1]$
P.R.4	$Exit = 0$; $rsubset(\ ) = C$; exit from procedure
P.R.5	if $r = 0$: $Exit = 1$; exit from procedure
P.R.6	$A = (S \sim C) \sim (C)_1$
P.R.7	if $A = \{\ \}$: $Exit = 1$; exit from procedure
P.R.8	$[C \sim (A)_1] \cup \{(A)_1\} \rightarrow C$
P.R.9	$b = r - |C|$
P.R.10	if $b \neq 0$: $C \cup (S \cap [(S)_b + 1]) \rightarrow C$
P.R.11	$Exit = 0$; $rsubset(\ ) = C$; exit from procedure

FIG. 5.17. A general permutation generation procedure.

m_i-combination of these positions is specified by T_i, $i = 0$ to $k-1$. Each time a T_i is chosen, the objects i are placed (line P.6) in positions p_j, $j \in T_i$; thus the resulting permutation is $p_{0 \text{ to } n-1}$. $S_{0 \text{ to } k-1}$ and $T_{0 \text{ to } k-1}$ are auxiliary arrays of sets used only in this generation. Their storage is not releasable (see § 2.4.2), hence they should be transmitted by name as arguments of the $gperm(\ )$ procedure if this procedure is referenced "simultaneously" from more than one place in a running program. The generation $rsubset(\ )$ is a direct generalization of the program of Fig. 5.4 to the case of r-subsets of an *arbitrary* set.

The program of Fig. 5.17 is a general-purpose procedure. If $k = 2$, then combinations are generated. If $k = n$, so that $m_i = 1$ for all i, then permutations of distinct objects are generated.

5.3.2. *Serial Number Calculation*

The program of Fig. 5.17 generates permutations (as vectors) in increasing lexicographic order. Given one of these permutations we may calculate its serial number from the progress of each of the k m_i-subset generations. For $i = 0$ to $k-1$, let

$$U_i = \binom{n - \sum_{j \in i}(m_j)}{m_i} \tag{5.9}$$

and for $i = 0$ to $k-2$, let

$$V_i = \prod_{j=i+1 \text{ to } k-1}(U_j). \tag{5.10}$$

The serial number N of a permutation $p_{0 \text{ to } n-1}$ is given by

$$N = \sum_{i=0 \text{ to } k-2}(M_i V_i), \quad 0 \leqslant M_i < U_i, \tag{5.11}$$

where each M_i is the serial number of an m_i-subset within its own generation. This process is essentially the calculation of a number from its mixed radix representation —see exercise 1 of § 3.1. Here U_i are the radices, V_i the associated products, and M_i the digits of the representation. A procedure to compute N is given in Fig. 5.18. This procedure is strongly dependent upon the objects being the natural numbers $0, 1, \ldots, k-1$.

The inverse of this function is given in Fig. 5.19. This is a straightforward conversion of the type discussed in § 3.1; the procedure *numtocomb*( ), which yields the set D, appears in Fig. 5.7.

As with other serial number conversions, this procedure can be used as a permutation generator by feeding it successive values of N—for example, it becomes a random permutation

N.0   *gpermtonum*(n,k,m,p)

N.1   for $i = 0$ to $k-1$: $H_i, M_i = 0$; $J_i = 1$

N.2   for $i = 0$ to $n-1$:

N.3 $M_{p_i} + \binom{H_{p_i}}{J_{p_i}} \rightarrow M_{p_i}$

N.4 $H_{p_i} + 1 \rightarrow H_{p_i}$; $J_{p_i} + 1 \rightarrow J_{p_i}$

N.5   for $q = 0$ to $p_i - 1$: $H_q + 1 \rightarrow H_q$

N.6 $V = 1$; $N = 0$; $q = m_{k-1}$

N.7   for $j = k-2, k-3, \ldots, 0$:

N.8 $N + M_j V \rightarrow N$; $q + m_j \rightarrow q$

N.9 $V\binom{q}{m_j} \rightarrow V$

N.10   *gpermtonum*( ) $= N$; exit from procedure

FIG. 5.18. Serial number calculation for permutations with repetition.

G.0   *numtogperm(N,n,k,"m")*

G.1 $V_k, V_{k-1} = 1; q = m_{k-1}$ "*V* is the array of (5.10)"

G.2   for $j = k-2, k-3, \ldots, 1$:

G.3 $q+m_j \to q; V_j = V_{j+1}\dbinom{q}{m_j}$

G.4 $S = $ (set) n

G.5   for $i = 0$ to $k-1$, as $q = n, n-m_{i-1}, \ldots$:

G.6   define Q, N by N/V_{i+1}

G.7 $D = numtocomb(Q,q,m_i)$

G.8   for $j \in S$, as $h = 0, 1, \ldots, q-1$:

G.9   if $h \in D$:

G.10 $p_j = i$

G.11 $S \sim \{j\} \to S$

G.12   *numtogperm*( ) $= p$; exit from procedure

FIG. 5.19. Conversion of a serial number to a general permutation.

generator by feeding it random values of N. As a generation procedure it is less efficient than *gperm*( ) of Fig. 5.17 but does contain only releasable storage, hence is useful in problems where many generations are occurring concurrently.

EXERCISES

1. Write a program to calculate the multinomial coefficient (5.7) from given values for k and $m_{0 \text{ to } k-1}$

2. (a) In what order does the program of Fig. 5.17 generate permutations of distinct objects when $k = n$?

(b) In what order does this program generate combinations when $k = 2$?

*3. Write a program to generate permutations with repeated objects by successive transposition. (Hint: Apply the generation suggested by exercise 7 of § 5.1.)

4. As noticed by Lehmer [144], adjacent transposition will not in general generate all permutations with repeated objects.

(a) Let permutations be vertices of a graph. Let an edge connect two vertices if and only if the permutations differ by a transposition of adjacent marks. Draw the graph associated with permutations of (0,0,1,1).

(b) Write a generation program where "most of the time" successive permutations differ by a transposition of adjacent marks.

5. Modify the algorithm of exercise 12 of § 5.2 to generate permutations with repeated objects.

5.4. Compositions

There are several diverse ways in which compositions may be generated. Each has its place in combinatorial computing.

5.4.1. *Relationship with Subsets—Serial Numbers*

The simplest, hence perhaps most important, generation schemes are those associated with the difference representation for compositions (see § 3.3.4). In that representation, a composition of n is expressed as a subset of $\{0,1,\ldots,n-2\}$, the parts being given essentially by differences of consecutive elements of the subset. A generation procedure of this type for all compositions is given in Fig. 5.20. This procedure yields the composition $(C_1, C_2, \ldots, C_m)$, where m, the number of parts, is produced as the value of the procedure.

C.0   *composition*$(n,\text{"}C\text{"})$

C.1   if *Entry* $= 0$: go to $\# 1$

C.2   for $S \subset (n-1)$:

C.3   for $a = -1, \ldots, s \to a, \ldots$, as $m = 1, 2, \ldots$, as $s \in S$:

C.4 $C_m = s - a$

C.5 $C_m = n - a - 1$

C.6   *composition*$(\ ) = m$; *Exit* $= 0$; exit from procedure

C.7 $\# 1$ reiterate

C.8   *Exit* $= 1$; exit from procedure

FIG. 5.20. Generation of compositions in difference representation.

The serial number of a composition S (expressed in the difference representation) is simply $N = nbr(S)$. Of course, $0 \leqslant N < 2^{n-1}$ since there are 2^{n-1} subsets of $\{0,1,\ldots,n-2\}$.

The association of subsets with compositions suggests generation schemes for various restricted classes of compositions. An m-part composition of n is associated with an $(m-1)$-subset of $\{0,1,\ldots,n-2\}$, hence one may generate m-part compositions by replacing line C.2 of Fig. 5.20 by

C.2   for *combination*$(n-1, m-1, \text{"}S\text{"})$:

(see Fig. 5.4) and including m as an argument in the heading at line C.0. This association also provides a serial number calculation for m-part compositions, by means of the *combtonum*$(\ )$ procedure of Fig. 5.7. (Note that there are $\binom{n-1}{m-1}$ m-part compositions.)

A slightly different association is involved in generating compositions into m parts *allowing null parts*—a generation which occurs in many enumerative calculations. Consider the bit-pattern 001101000111 which is associated in the difference representation with the 7-part composition of 13 (1,1,1,4,2,1,3). Subtract one from each part. This yields (0,0,0,3,1,0,2), which is a composition of 6 into 7 parts allowing parts equal to zero. In this way we can make a one-to-one association between an m-part composition of n allowing zeros and an m-part composition of $n+m$ into positive parts. Computationally the generation of these null-part compositions merely involves an alternate interpretation of the bit-pattern representation of an $(m-1)$-subset of $\{0,1,\ldots,n+m-2\}$. Specifically, the parts are given by the number of 0-bits between successive 1-bits of the pattern. The serial number $N, 0 \leqslant N < \binom{n+m-1}{m-1}$, of such a composition can be calculated, as before, from the *combtonum*( ) procedure.

5.4.2. *Generation in Signature Form*

As shown in § 4.1.1, the generation of a signature in which the components are selected one at a time to satisfy a given set of restrictions may conveniently be written as a nest of iterations. The nest index specifies the component positions while the general loop index specifies the component values; that is, the nest vector *is* the signature. Figure 4.4 shows a simple nest which generates m-part compositions of n.

In general, quite arbitrary restrictions on the parts of a composition may be accommodated with modifications upon the index range of the iteration quantifier. As an example we present in Fig. 5.21 a procedure which generates m-part compositions of n subject to the restriction on each part C_i that $p \leqslant C_i \leqslant P$. In this program (as in Fig. 4.4), the quantity s is the running sum of parts. The component upper limit $n+i-p(m+s)$ insures that C_i does not become so large that there is too little of n left for distribution among the remaining parts, while the lower limit $n-s-P(m-i)$ insures that C_i is large enough to allow each remaining part to stay $\leqslant P$.

C.0   *composition(n,m,p,P,"C")*

C.1   if *Entry* = 0: go to # 1

C.2   with $i = 1$ to $m-1$, as $s = 0, 0+C_{i-1}, \ldots$:

C.3   · ·
                         ·

C.4   for $C_i = max[p, n-s-P(m-i)]$ to $min(P, n+i-p(m+s))$:

C.5   · ·
                         ·

C.6 $C_m = n-s$; *Exit* = 0; exit from procedure

C.7   # 1 $i = m$; $s = n-C_m$

C.8   *Exit* = 1; exit from procedure

FIG. 5.21. A restricted m-part composition generation procedure.

5.4.3. *Compositions as Permutations of Partitions*

A third means of composition generation is illustrated in Fig. 5.22. This scheme permutes each partition of n in every possible way, thus forming all compositions of n.

The *partition*( ) generation procedure (§ 5.5) produces in turn, at line C.2, the partitions $(P_1, P_2, \ldots, P_{n*})$. A partition is analyzed for object repetition at lines C.3 through C.6. The specification $(m_0, m_1, \ldots, m_{k-1})$ is then used as input to the *gperm*( ) procedure (§ 5.3) which produces in turn, at line C.7, each permutation. The composition itself is formed at line C.8 from this permutation and the vector of partition parts $(V_0, V_1, \ldots, V_{k-1})$.

C.0   *composition*(n;"C")

C.1   if *Entry* $= 0$: go to # 1

C.2   for each *partition*(n;"P","$n*$")

C.3 $m_0 = 1; \; k = 1; \; V_0 = P_1$

C.4   for $i = 2$ to $n*$:

C.5   if $P_i = P_{i-1}$: $m_{k-1} + 1 \to m_{k-1}$

C.6   otherwise: $m_k = 1; \; V_k = P_i; \; k+1 \to k$

C.7   for each *gperm*($n*,k$,"m";"p")

C.8   for $i = 0$ to $n* - 1$: $C_{i+1} = V_{p_i}$

C.9   *Exit* $= 0$; exit from procedure

C.10   # 1   reiterate

C.11   *Exit* $= 1$; exit from procedure

FIG. 5.22. Generation of compositions as permutations of partitions.

EXERCISES

1. Write an expression, or program, for the serial number of a composition $(C_1, C_2, \ldots, C_m)$ of n into positive parts.

2. Write a program which generates random compositions of n.

3. Exhibit the one-to-one correspondence between compositions of six into three positive parts and compositions of two into four nonnegative parts which is indicated by the bit-pattern representation of such compositions.

4. Write a program to calculate a composition of n into m nonnegative parts as a vector $(c_1, c_2, \ldots, c_m)$ from the $(m-1)$-subset of $n+m-1$ called S.

5. Write a procedure which generates compositions of n into odd parts.

6. Modify the procedure of Fig. 5.22 so that an input toggle y indicates whether the remaining compositions associated with the current partition need be generated or not.

7. Devise a serial number scheme which assigns numbers in increasing order to compositions as generated by the procedure of Fig. 5.22. Assume a partition serial number scheme exists—see § 5.5.2.

5.5. Partitions

In this section we discuss generation schemes for partitions of a whole number and for partitions of a set. These generation schemes are required in many combinatorial investigations, often as a basis for more involved generations (e.g. see § 5.4.3). Since numerical partitions naturally arise in many classification schemes (a simple example is the classification of group elements by cycle structure), their serial number calculation (§ 5.5.2) often appears as one step in an involved indexing scheme. Such an application appears in the incidence matrix invariant calculation which is discussed in § 7.1.1. An application of the special set partition generation scheme of § 5.5.4 appears in the four-color problem discussion of § 7.3.

5.5.1. *Numerical Partitions in Signature Form*

The partitions of a whole number required for a particular study are frequently subject to certain restrictions. Thus the most important partition generation algorithm is the signature generation since it is easily modified to handle arbitrary restrictions. In fact, even though an alternate representation (e.g. the rim representation) is more suited to a given calculation, it is often preferable to generate the partitions as signatures and then convert to the representation appropriate for manipulation or storage.

A procedure for the signature generation is given in Fig. 5.23. It produces the partitions $(P_1, P_2, \ldots, P_m)$ as vectors in decreasing order: the first partition formed is (n), the second is $(n-1, 1)$, and so forth; the last one formed is $(1, 1, \ldots, 1)$. Given a partition the next partition is formed by subtracting one from the rightmost part P_k, which is greater than one, and then distributing what is left of n—namely $N = \sum_{k=1 \text{ to } k}(P_h)$—as "quickly as possible" (see line P.5), subject to the inequality restrictions, among the parts P_{k+1}, $P_{k+2}, \ldots$. Upon exit from the procedure the index m supplies the number of parts of the partition.

If we wish to generate partitions with parts $\leqslant M$, then M (an additional input argument for the procedure) becomes an argument of the *min*( ) function. To generate partitions with exactly m^* parts the quantifier at line P.5 is replaced by the line for $P_{m+1} = min(P^m, N, N-m^*+m+1), \ldots, P_{m+1}-1 \rightarrow P_{m+1}, \ldots$, until $P_{m+1} < N/(n^*-m)$. Here the initial value for P_{m+1} has the additional upper limit $N-m^*+m+1$ to insure that there is enough of n left for distribution among the remaining m^*-m-1 parts. The last value of P_{m+1} is the smallest integer which is not less than $N/(n^*-m)$; this prevents assignment of values for P_{m+1} which leave so much of n undistributed that the inequalities $P_1 \geqslant P_2 \geqslant \ldots \geqslant P_{m*}$ cannot be satisfied.

P.0 $partition(n;"P","m")$

P.1   if $Entry = 0$: go to $\# 1$

P.2 $P_0 = n$

P.3   with $m = 0, 1, \ldots,$ as $N = n, n-P_m, \ldots,$ until $N = 0$:

P.4 $\cdot\ \cdot\ .$

P.5   for $P_{m+1} = min(P_m, N), \ldots, P_{m+1}-1 \rightarrow P_{m+1}, \ldots,$

                            until $P_{m+1} < 1$:

P.6 $\cdot\ \cdot\ .$

P.7 $Exit = 0$; exit from procedure

P.8 $\# 1$ $N = 0$

P.9 $Exit = 1$; exit from procedure

FIG. 5.23. Basic recursive generation of partitions.

5.5.2. Serial Numbers for Numerical Partitions

The number of partitions of n into parts less than or equal to m, $R_{n,m}$, may be calculated by the recursion shown in Fig. 5.24. That the recurrence relation of line Q.5 is true may be seen classifying the partitions into those with largest part exactly equal to m (the second summand) and those with largest part less than m (the first summand).

Using these numbers we may establish a one-to-one correspondence between the unrestricted partitions of n and the numbers $0, 1, \ldots, R_{n,n}-1$ (note that $R_{n,n}$ is the number of unrestricted partitions of n). This correspondence is given by the procedure *partonum( )* and its inverse *numtopart( )* presented in Fig. 5.25. From a partition $(P_1, P_2, \ldots, P_m)$ and the precomputed 2-array R, the procedure *partonum( )* calculates the number of partitions which follow the given partition in the sequence produced by *partition( )* of Fig. 5.23. Thus $(1,1,\ldots,1)$ is associated with 0, $(2,1,\ldots,1)$ with 1, $\ldots$, (n) is associated with $R_{n,n}-1$. The procedure *numtopart( )* reverses the process; it is a straightforward conversion of N to $(P_1, P_2, \ldots, P_m)$, analogous to *numtoperm( )* of Fig. 5.12. The index m supplies, upon exit from the procedure, the number of parts of the partition.

Q.1 $R_{0,0} = 1$

Q.2   for $n = 1, 2, \ldots, Limit$:

Q.3 $R_{n,0} = 0$

Q.4   for $m = 1, 2, \ldots, n$:

Q.5 $R_{n,m} = R_{n,m-1} + R_{n-m,min(n-m,m)}$

FIG. 5.24. Calculation of number of partitions of n into parts $\leqslant m$.

N.0 $partonum(n,\text{"}P\text{"},\text{"}R\text{"})$

N.1 $partonum(\ ) =$

N.2 $\sum_{h=1,2,\ldots,\text{ as } k=n,\ldots,k-P_{h-1}\to k,\ldots,\text{ until } k=0}\left(R_{k,P_h-1}\right)$

N.3   exit from procedure

P.0 $numtopart(n,N,\text{"}R\text{"};\ \text{"}P\text{"},\text{"}m\text{"})$

P.1   for $m = 0, 1, \ldots,$ until $n = 0$:

P.2 $P_{m+1} = (\text{first } k = n-1, n-2, \ldots \ni R_{n,k} \leqslant N)+1$

P.3 $N-R_{n,k} \to N;\ n-P_{m+1} \to n$

P.4   exit from procedure

FIG. 5.25. Conversions between partitions and their serial numbers.

5.5.3. *Partitions of a Set*

Consider a partition $P_{1\text{ to }r}$ of the set $\{0,1,\ldots,n-1\}$; that is, $P_i \subset n$, $P_i \neq 0$, $P_i \cap P_j = 0$ for $i \neq j$, $\bigcup_{i=1\text{ to }r}(P_i) = n$, and by convention $(P_1)_1 < (P_2)_1 < \ldots < (P_r)_1$. The array $P_{2\text{ to }r}$ is a partition of $n \sim P_1$, the array $P_{3\text{ to }r}$ is a partition of $n \sim (P_1 \cup P_2)$, and so forth. We may therefore generate partitions recursively as shown in Fig. 5.26, the part P_i being chosen from the set $S_i = n \sim \bigcup_{j=1\text{ to }i-1}(P_j)$. To preserve the ordering of the parts in the representation, P_i must contain the smallest element of S_i, say a_i. In forming all partitions, P_i will eventually contain in turn each subset p_i of $S_i \sim \{a_i\}$—these subsets are formed by the quantifier at line P.5. When all elements of $\{0,1,\ldots,n-1\}$ have been distributed, i.e. when S_i is void, we have a partition $P_{1\text{ to }i-1}$ (line P.9).

Consider now the first level of this process and imagine the subsets of $n \sim \{0\}$ classified by their cardinality. There are $\binom{n-1}{k}$ subsets with cardinality k, and for each

P.1 $S_1 = (\text{set})\ n$

P.2   with $i = 1, 2, \ldots,$ until $S_i = \{\ \}$:

P.3 $\therefore$

P.4 $a_i = (S_i)_1$

P.5   for $p_i \subset S_i \sim \{a_i\}$:

P.6 $P_i = p_i \cup \{a_i\}$

P.7 $S_{i+1} = S_i \sim P_i$

P.8 $\therefore$

P.9   USE PARTITION $P_{1\text{ to }i-1}$

FIG. 5.26. Basic recursive generation of partitions of a set.

such subset one generates all partitions of a set with $n-1-k$ elements. Summing over k we see that there are altogether

$$B_n = \sum_{k=0 \text{ to } n-1} \left[\binom{n-1}{k} B_{n-1-k} \right] \tag{5.12}$$

partitions of a set with n elements. (This recurrence relation is identical to (3.20) under the index transformation $n-1-k \to k$.) To use this formula it is necessary to have $B_0 = 1$; then $B_1 = 1$, $B_2 = 2$, $B_3 = 5$, $B_4 = 15$, and so forth—see Appendix I.

Using the ordering indicated by this derivation we may assign a serial number N to a partition $P_{1 \text{ to } r}$ by means of the formula

$$N = \sum_{i=1 \text{ to } r} \left(\sum_{k=0 \text{ to } |P_i|-2} \left[\binom{|S_i|-1}{k} B_{|S_i|-1-k} \right] \right.$$
$$\left. + \sum_{j=2 \text{ to } |P_i|} \left[\binom{|(P_i)_j \cap S_i|-1}{j-1} \right] B_{|S_i|-|P_i|} \right). \tag{5.13}$$

In this formula, as in Fig. 5.26, $S_i = n \sim \bigcup_{j=1 \text{ to } i-1}(P_j)$.

Though complicated in appearance, this formula is a straightforward application of the recursive generation just described. The first term inside the parentheses (i.e. the second summation) counts the partitions "preceding" (when the subsets are generated in increasing order of their cardinality) the given partition which have initial part of smaller cardinality. The second term of the parentheses counts those partitions preceding the given partition which have an initial part different from that of the given partition but with the same cardinality. [The first factor of this second term of the parentheses (i.e. the third summation) gives the serial number of P_i within S_i (see exercise 8 of § 5.1) subject to $(S_i)_1 \in P_i$, of course.] The serial number is then formed as the summation (the first one of (5.13)) of these two counts over each recursive step, i.e. over each partition part. Note that the procedure of Fig. 5.26 produces partitions in increasing order of this serial number only if the quantifier of line P.5 produces subsets in increasing order of their cardinality (see exercise 14 of § 5.1). As an example, Fig. 5.27 lists the fifteen partitions of $\{0,1,2,3\}$ in order of their serial numbers.

N	P	N	P
0	{0} {1} {2} {3}	9	{0,3} {1} {2}
1	{0} {1} {2,3}	10	{0,3} {1,2}
2	{0} {1,2} {3}		
3	{0} {1,3} {2}	11	{0,1,2} {3}
4	{0} {1,2,3}		
		12	{0,1,3} {2}
5	{0,1} {2} {3}		
6	{0,1} {2,3}	13	{0,2,3} {1}
7	{0,2} {1} {3}	14	{0,1,2,3}
8	{0,2} {1,3}		

FIG. 5.27. Serial number assignment for partitions of a 4-set.

5.5.4. *Binary Trees—Partitions Without Crossing*

We conclude this chapter with a discussion of the generation of an important [138] class of binary trees—search trees with a fixed number of links and the out-valence of every node either zero or two. Since each nonterminal node has precisely two out-going links, the number of links is even, say $2n$; then the number of nonterminal nodes is n. Since each nonterminal node, except the origin, is incident with three edges and each edge has two nodes, there are $n+1$ terminal nodes and $2n+1$ nodes in all.

It can be shown (the derivation is sketched later in this section) that the number of such trees is $\dfrac{1}{n+1}\dbinom{2n}{n}$. A short table of these important numbers is given in Appendix I. For reference purposes we refer to these numbers as *Segner numbers*[†] after J. A. von Segner who first developed a recurrence relation for their calculation (see [136]).

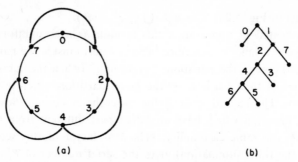

(a)   (b)

FIG. 5.28. Association of certain set partitions to binary trees.

Now consider a special class K of partitions of $\{0,1,\ldots,2n-1\}$. A partition $Q_{0\text{ to }r-1}$ is in K if it satisfies the requirements (1) $Q_i \subset \{0,2,4,\ldots,2n-2\}$ or $Q_i \subset \{1,3,5,\ldots, 2n-1\}$ for each $i \in r$, (2) either $S_i \subset S_j$, $S_j \subset S_i$, or $S_j \cap S_i = 0$ for all i and j, where S_k, $k \in r$, denotes the set $(Q_k)^1 \sim (Q_k)_1$, and (3) no partition $P_{0\text{ to }s-1}$ exists, also satisfying (1) and (2), such that $s < r$ and $\forall_{j\in r}(\exists_{i\in s}(Q_j \subset P_i))$ is true. Restriction (2) says roughly that the parts do not "cross" while restriction (3) insures that the partitions are in some sense "maximal".

The partition $Q_0 = \{0\}, Q_1 = \{1,7\}, Q_2 = \{2,4,6\}, Q_3 = \{3\}, Q_4 = \{5\}$ of $\{0,1,\ldots,7\}$ is an example of such a partition. It may be represented pictorially as shown in Fig. 5.28(a). Furthermore, one may establish a one-to-one correspondence between these restricted partitions and $2n$ edge binary trees as illustrated in Fig. 5.28(b). Left-falling links are labeled with even numbers and right-falling links with odd numbers. The labeling begins at the left link emanating from the origin and proceeds as a systematic tree scan alternately assigning even and odd numbers in winding down through the tree. Parts of the partition are given by straight lines of adjacent links. We see from this association that

[†] These numbers are referred to as the *Catalan numbers* by some authors.

the number of parts of the partition equals the number of terminal nodes of the tree; therefore, $r = n+1$ and we are concerned with partitions $Q_{0 \text{ to } n}$.

This association between certain noncrossing set partitions and binary trees is just one of many related associations concerning binary trees. The problem considered by Segner was the enumeration of the dissections of a labeled polygon into triangles by means of nonintersecting diagonals—there is a one-to-one correspondence between such dissections of an n-gon and $2n-4$ edge binary trees. Another example, more pertinent here, is the existence of an association between $2n$ edge binary trees and partitions of $\{0,1,\ldots,n-1\}$ which do not cross (not necessarily maximal). This correspondence may be established using a normal scan labeling of the nonterminal nodes; the subsets correspond to right-falling lines. Of course, whenever such an association exists, the objects are enumerated by the Segner numbers.

A recurrence relation for enumerating binary trees (hence for Segner numbers) is derived as follows. Let T_n be the number of binary trees with $2n$ edges. We consider an isolated vertex to be the unique tree for $n = 0$; thus, $T_0 = 1$. Also, $T_1 = 1$, the unique

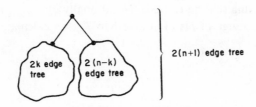

FIG. 5.29. The basic recursive step for binary tree enumeration.

tree for $n = 1$ has two links and two terminal nodes. One can form trees with $2(n+1)$ edges, $n \geqslant 1$, by attaching trees with $2k$ edges and $2(n-k)$ edges respectively on the two terminal nodes of this 2-edge tree—see Fig. 5.29. For a particular k, this yields $T_k \times T_{n-k}$ distinct trees. Summing over k we get

$$T_{n+1} = \sum_{k=0 \text{ to } n}(T_k T_{n-k}) \tag{5.14}$$

which is valid for $n \geqslant 0$. Using "generating functions", the binomial theorem and (5.14) it is not difficult to verify (see Netto [32] or Hall [20], and exercise 10) the formula for Segner numbers

$$T_n = \frac{1}{n+1}\binom{2n}{n} \tag{5.15}$$

stated previously.

The recursive process used to derive (5.14) may be employed to generate $2n$ edge binary trees or in particular objects in one-to-one correspondence with such trees. For small n, say $n \leqslant 6$, the recursion may be performed straightforwardly as we may save all the smaller objects for use in forming the larger ones. Such a generation is exhibited in Fig. 5.30. This program constructs the partitions-without-crossing discussed earlier

$$q_{0,1,0} = \{ \; \}$$

for $m = 1$ to n:

$\qquad j = 1$

$\qquad$ for $x = 0$ to $m-1$, as $y = m-1, m-2, \ldots, 0$:

$\qquad\qquad$ for $p = 1$ to T_x; for $v = 1$ to T_y:

$\qquad\qquad\qquad$ for $k = 0$ to x: $q_{m,j,k} = q_{x,p,k}$

$\qquad\qquad\qquad$ for $k = 0$ to y: $q_{m,j,k+x+1} = q_{y,v,k} \oplus \{2x+2\}$

$\qquad\qquad\qquad q_{m,j,0} \cup \{2x\} \rightarrow q_{m,j,0}$

$\qquad\qquad\qquad q_{m,j,x+y+1} \cup \{2x+1\} \rightarrow q_{m,j,x+y+1}$

$\qquad\qquad\qquad j+1 \rightarrow j$

FIG. 5.30. "Binary tree" generation.

(a variant of the labeling scheme is used). Notationally $q_{m,j,0 \text{ to } m}$ is the jth partition of $\{0,1,\ldots,2m-1\}$. The range of j is of course 1 to T_m; we calculate the array for $m = 0$ to n.

EXERCISES

1. Write a procedure which generates partitions directly in the rim representation (see § 3.3.4).

2. Write programs analogous to *partonum*( ) and *numtopart*( ) of Fig. 5.25 for partitions of n into exactly m parts.

3. Write a procedure to generate all partitions which are a "refinement" of a given partition. The partition $(q_1, q_2, \ldots, q_r)$ is a *refinement* of $(p_1, p_2, \ldots, p_s)$ when there exists a partition $Q_{1 \text{ to } s}$ of the set $\{1, 2, \ldots, r\}$ and a permutation $f_{1 \text{ to } s}$ of $(1, 2, \ldots, s)$ such that $p_{f_k} = \bigcup_{j \in Q_k}(q_j)$ for each $k = 1$ to s.

4. Combining permutation generation (e.g. Fig. 5.17) with partition generation, write a procedure which generates permutations of n distinct objects in cycle form by "conjugate class"; that is, all permutations with the same cycle structure are to be generated sequentially.

5. A d-dimensional numerical partition may be defined as a set S of d-tuples of natural numbers such that $(x_1, x_2, \ldots, x_d) \in S \Rightarrow (x_1^*, x_2^*, \ldots, x_d^*) \in S$ whenever $0 \leqslant x_i^* \leqslant x_i$ for $i = 1, 2, \ldots, d$. Thus an ordinary partition is two-dimensional, the pairs being the coordinates of the nodes in the Ferrers graph of the partition (see § 3.3.4). Write a program to generate d-dimensional partitions of n (see [135]).

***6.** Invert formula (5.13). That is, write a program to compute the partition of $\{0,1,\ldots,n-1\}$ associated with an arbitrary natural number N, $N < B_n$.

7. List the fifteen partitions of $\{0,1,2,3\}$ in the order produced by the program of Fig. 5.26 when the subsets p_i are generated at line P.5 in increasing order of $nbr(p_i)$.

8. List all fourteen partitions of $\{0,1,\ldots,7\}$ of the type illustrated in Fig. 5.28.

9. Find a one-to-one correspondence between dissections of a labeled n-gon into triangles by $n-3$ non-intersecting diagonals and $2n-4$ edge binary trees.

10. Write $f(x) = T_0 + T_1 x + T_2 x^2 + \ldots$. Form $[f(x)]^2$. Using (5.14), find a quadratic equation which $f(x)$ satisfies. Solving this equation and expanding the result by means of the binomial formula, prove (5.15).

11. Show that the number of ways of forming the product $a_1 a_2 \ldots a_n$ (in that order) by using a binary but nonassociative "multiply" is $(2n-2)!/[n!(n-1)!]$. For example, $a_1 a_2 a_3 a_4$ can be formed as $((a_1 a_2)(a_3 a_4))$, $(((a_1 a_2)a_3)a_4)$, $((a_1(a_2 a_3))a_4)$, $(a_1((a_2 a_3)a_4))$, and $(a_1(a_2(a_3 a_4)))$.

12. Write a backtrack program which generates binary trees.

13. Write a program which calculates the serial number of an arbitrary partition of $\{0, 1, \ldots, 2n-1\}$ of the type generated by the program of Fig. 5.30.

*14. For a partition $p_{1 \text{ to } k}$ of $2n$, let $N(p)$ be the number of distinct ways in which $2n$ objects of specification p can be placed in n indistinguishable boxes, each box containing exactly two objects. For instance, for $n = 5$, and $p = (6,4)$, $N(p) = 3$, since objects 000000 and 1111 may be placed as [00][00][00][11][11], [00][00][01][01][11], or [00][01][01][01][01]. Write a program to compute $N(p)$ for an arbitrary given partition of an even number.

15. For $n = 12$ to 30, calculate the number of vectors $(a_5, a_6, a_7, \ldots)$ whose nonnegative components satisfy the conditions

$$a_5 + a_6 + \ldots + a_k + \ldots = n$$

and

$$a_7 + 2a_8 + \ldots + (k-6)a_k + \ldots = a_5 - 12.$$

16. Show that the number of m-part partitions of $\{1, 2, \ldots, n\}$ is the Stirling number of the second kind, $S_{n,m} = S_{n-1,m-1} + mS_{n-1,m}$ (see Appendix I). Can a serial number scheme for partitions of a set be based on the precalculation of these or related numbers?

17. Write a program to generate r-combinations of n not necessarily distinct objects, say, of n objects of specification m, where m is a partition of n.

*18. Let $p_{1 \text{ to } j}$ and $q_{1 \text{ to } k}$ be arbitrary partitions of n. Consider a store of n labels: p_1 ones, p_2 twos, $\ldots, p_j$ j's. Write a program to calculate $y_{p,q}$, the number of ways in which these n labels can be attached to the nodes of the Ferrers graph (§ 3.3.4) associated with q in such a way that the labels have nondecreasing values from left to right in the rows and strictly increasing values from top to bottom in the columns. (Note: When $p = (1, 1, \ldots, 1)$, these figures are called *standard Young tableaux* [87]. How might your approach change if a complete table, i.e. for all p and q, were desired in place of an isolated value?)

19. Devise an efficient serial number indexing scheme for numerical partitions of n into m or fewer parts. (Hint: There is a one-to-one correspondence between these partitions and those of n into parts less than or equal to m—consider transpositions of the Ferrers diagram.)

20. Show that the total number of m-part numerical partitions with part size 0 to m inclusive is $\binom{n+m}{n}$. Devise a serial number scheme for these partitions (see § 7.1.1).

CHAPTER 6

ADDITIONAL BASIC TECHNIQUES
AND MANIPULATIONS

THIS chapter presents a general discussion of further elementary techniques of combinatorial computing. The major sections are essentially independent and may be read in arbitrary order.

6.1. Sieving Processes

A common method of producing a desired set of objects is to eliminate unwanted objects from a universe known to contain the desired set as a subset. We call such a process in which several objects are eliminated in each of a series of well-defined steps a *sieving process* or, more briefly, a *sieve*. For our purposes the universe for a sieve is finite. Therefore, since we require that each step eliminate at least one object, the number of steps needed to produce an answer, although possibly unknown, is also finite. A sieve may thus be visualized as a process which calculates, successively, n sets c_1, $c_2, \ldots, c_n$ of unwanted objects in order to produce a desired subset

$$s_n = s_0 \sim \bigcup_{i=1 \text{ to } n} (c_i) \tag{6.1}$$

of a given universe s_0. Alternatively, s_n may be calculated from the recursion

$$\text{for } i = 1, 2, \ldots, n: s_i = s_{i-1} \sim c_i \tag{6.2}$$

rather than at one fell swoop as indicated by (6.1).

This section discusses "general-purpose" computer implementation of special classes of sieves. (See [173] for an announcement of a "special-purpose" sieving machine.) In most cases, a sieving process is effective when the "culls" (i.e. the unwanted objects) naturally group themselves into few, easily calculable subsets.

6.1.1. *Modular Sieves*

A class of sieves useful in many number-theoretic investigations and which contains the sieve of Eratosthenes and the Chinese remainder sieve (i.e. the conversion from Chinese representation—see § 3.1.3) as special cases comprises the *modular sieves*. For such a sieve one is given (1) n pairwise relatively prime integers m_1, m_2, ..., m_n—the *moduli*, (2) n subsets $R_i \subset m_i$, $i = 1$ to n—the *residue sets*, and (3) a set of consecutive integers—the *universe*. The culls c_i are those integers not representable by the form $km_i + (R_i)_j$ for some k and j. For example, in the sieve of Eratosthenes, the moduli m_i, $i \in m$, are the prime numbers less than (a given) b, the residue set R_i equals $m_i \sim \{0\}$, and the universe s_0 equals $\{b, b+1, \ldots, b^2\}$. The cull c_i consists simply of multiples of m_i. The answer sought is the set of primes in s_0.

D. H. Lehmer [174] was the first to consider the execution of these sieves on a digital computer. They are readily adaptable to machine execution since the culls associated with each linear form (each i) are simply "translated" multiples of m_i. The basic program structure for executing a modular sieve is illustrated in Fig. 6.1; for simplicity we have

M.1   for $i = 1$ to n:

M.2 $C = m_i \sim R_i$

M.3 $c_i = \{\ \}$

M.4   for $k = 0, 0+m_i, \ldots,$ until $k > K$:

M.5 $c_i \cup (C \oplus \{k\}) \rightarrow c_i$

M.6 $s_i = s_{i-1} \sim c_i$

M.7   "At this point we have the answer s_n."

FIG. 6.1. A basic modular sieve program.

assumed the universe s_0 to be $\{0, 1, \ldots, K\}$. Often in practice the c_i are not explicitly calculated and s is not subscripted; for instance, line M.5 is replaced by

$$s \sim (C \oplus \{k\}) \rightarrow s$$

and lines M.3 and M.6 are omitted.

When using this program for the sieve of Eratosthenes, it is customary (for sake of compactness) to pre-sieve with the prime 2, letting $m_1 = 3$, $m_2 = 5$, etc., and letting $j \in s_0$ represent the integer $2j+1$. Note that for this sieve the set C (defined at line M.2) is simply $\{0\}$. The Chinese remainder sieve occurs when $s_0 = \{0, 1, \ldots, \prod_i (m_i) - 1\}$ for given relatively prime integers m_1, m_2, ..., m_n, and R_i is a singleton set containing a given residue modulo m_i. The answer set s_i contains but a single element, the unique (by the Chinese remainder theorem) solution to the set of congruences

$$x \equiv (R_i)_1 (\text{mod } m_i), \quad i = 1 \text{ to } n,$$

subject to the condition $0 \leqslant x < m_1 m_2 \ldots m_n$.

Of course, the program of Fig. 6.1 may be recast in many ways depending on the special sieve and particular computer system involved. A common modification, used both to take advantage of the efficiency of the $\oplus$ operation when the sets are represented

S.1 $P = \prod_{i=1 \text{ to } n}(m_i)$

S.2 $M = \bigcup_{i=1 \text{ to } n;\ k=0,\ 0+m_i,\ \dots,\ \text{until } k=P}[(m_i \sim R_i) \oplus \{k\}]$

S.3   "M consists of culls in $\{0,1,\dots,P-1\}$."

S.4 $S = b \sim a$

S.5   for $k = P[a/P],\ \dots,\ k+P \to k,\ \dots,$ until $k \geqslant b$:

S.6 $S \sim (M \oplus \{k\}) \to S$

S.7   "The sieving process is complete; S is the answer."

FIG. 6.2. A modified modular sieve program.

by bit-patterns and to eliminate essentially repetitive computation when $\prod_i(m_i)$ is small compared to the universe, is presented in Fig. 6.2. First, the set of culls in the interval $\{0,1,\dots,\prod_i(m_i)-1\}$ is computed (line S.2). This set is then used as a stencil in end-to-end positions across the universe $\{a,a+1,\dots,b-1\}$ to determine and eliminate larger culls (lines S.5 and S.6).

6.1.2. *Isomorph Rejection by Sieve*

In the modular sieves, the objects being sifted are the integers themselves; an element of s_0 for the computer program represents itself (or perhaps a simple linear function of itself, as with the most common version of the sieve of Eratosthenes). In general, the computerized s_0 is merely a mask whose integral elements index other more complex objects; the rules which determine cull objects may be quite arbitrary.

In combinatorial computing it is frequently necessary to extract from a given set of objects a subset of objects which are pairwise "nonisomorphic" (i.e. inequivalent) under some prescribed rule of equivalence. Such a problem is often conveniently solved by a sieving process.

Suppose there are n objects not all pairwise nonisomorphic. We first label the objects (usually not arbitrarily) with the indices 0, 1, $\dots$, $n-1$ (see § 3.2.1). We then write a generation procedure, say *isomorph(i)*, which successively calculates the indices corresponding to objects isomorphic, but not identical, to the object with index i. The straightforward sieving process given in Fig. 6.3 will then produce a mask M indicating a set of pairwise nonisomorphic objects—a set of representatives, one representative from

$M = (\text{set})\ n$

for $i \in M$:

    for each *isomorph(i)* $\to j: M \sim \{j\} \to M$

"At this point M is an answer."

FIG. 6.3. A bookkeeping program for isomorph rejection by sieve.

each equivalence class. Of course, the precise representatives produced depends on the order bestowed upon the objects by the labeling scheme.

The queens' problem mentioned in exercise 1 of § 5.2 affords a reasonably simple example of isomorph rejection by sieve. Suppose that one has available *all n* solutions to the problem for an $N \times N$ board, and wishes to produce solutions inequivalent under arbitrary rotation and reflection of the chessboard. A solution to the queens' problem is a permutation of $\{0,1,\ldots,N-1\}$, and it is convenient to label solutions with the serial numbers of the corresponding permutations. Let these serial numbers for the solution-permutations be the entries of the array $r_{0 \text{ to } n-1}$. For an index i, $i \in n$, the procedure *isomorph(i)* first computes the permutation $p_{0 \text{ to } N-1}$ associated with r_i by means of the *numtoperm( )* procedure (Fig. 5.12). It then generates each permutation equivalent to p under symmetries of the square (§ 6.5.2). Each of these permutations is then converted by a *reverse process* into a cull index for use by the sieve of Fig. 6.3. This reverse process, which is part of the *isomorph( )* procedure, consists of two steps: (1) computation of the serial number for a permutation by means of the *permtonum( )* procedure of Fig. 5.12, and (2) location of that serial number in the list $r_{0 \text{ to } n-1}$ by means of the binary search of Fig. 3.3.

Special attention should be paid to the role of permutation ranking in this example. Note that if N were reasonably small the set $\{0,1,\ldots,N!-1\}$ could be used as a basic universe for the sieving process and the *isomorph( )* procedure would be simpler since the array r, hence the initial table look-up and the binary search, would not be needed. In general one is confronted with this choice between use of a small universe resulting in a complex *isomorph( )* procedure and use of a large universe yielding a simple procedure. Additional discussion of isomorph rejection occurs in §§ 6.6 and 7.1.

6.1.3. *Further Examples of Recursive Sieving*

In the execution of a modular sieve as exemplified by the program of Fig. 6.1 it is assumed that the set of moduli is predetermined; the order in which the indexing is specified (line M.1) is of no consequence. On the other hand, in the execution of a rejection sieve—Fig. 6.3—the sequence of values assigned to the loop index is in fact only determined as the calculation proceeds, since elements of the index set M ahead of the current value of i are being eliminated within the loop. In this case, the index sequence actually yields the desired answers. Many sieving processes have this recursive property; in fact, the sieve of Eratosthenes itself can be so constructed.

An illustration of a recursive sieve, whose character is rather different from the isomorph rejection process, is given in Fig. 6.4. This program constructs a subset of the odd prime numbers with the property that every nonnegative even number $< 2N$ is the sum of two members of the subset. Here P is the (given) set of odd primes indexed such that $j \in P$ implies $2j+1$ is prime (i.e. $P = \{1,2,3,5,6,8,\ldots\}$), S is the answer subset indexed as is P, and E is the set of even numbers indexed such that $n \in E$ implies $2n$ is even. The program (1) finds the first even number n not yet represented as the sum of primes (line

G.1 $S = \{1\}$

G.2 $E = N \sim \{0,1,2\}$ "Begin by representing the even number 6"

G.3   for $n \in E$:

G.4 $p = [(\{n\} \ominus S) \cap P]^1$

G.5   if nonexistent: stop

G.6 $S \cup \{p\} \rightarrow S$

G.7 $E \sim (S \oplus \{p\}) \rightarrow E$

G.8   "When we reach this point, S is the desired subset."

FIG. 6.4. Example of a recursive sieve.

G.3), (2) finds if possible (line G.5) a "large" prime p whose inclusion in S allows representation of n (line G.4), adjoins this prime to S (line G.6), and then (3) eliminates those even numbers which are of the form $s+p$ for $s \in S$ (line G.7). Sieves of this sort have been used by Shen to check the Goldbach conjecture (i.e. every even integer > 4 is the sum of two odd primes) for even numbers $< 33,000,000$ [185] and by Stein and Stein to check the conjecture up to 10^8 as well as for numerous other Goldbach studies [187, 188].

Another example of a sieve which is inherently recursive is given by the generation of "lucky numbers". As defined by Ulam *et al.* [161], a *lucky number* is a positive integer which survives the systematic elimination defined as follows: From the list of integers 1, 2, 3, 4, ..., eliminate every second member, leaving the odd numbers 1, 3, 5, ... Now strike out every third member of this new list, leaving 1, 3, 7, 9, 13, 15, 19, 21 ... From this new list eliminate every *seventh* member (since, on this third step, 7 is the third survivor—7 is also the first integer of the list other than unity which has not yet been used as a "sieving number"), leaving 1, 3, 7, 9, 13, 15, 21, ...; and so forth. A straightforward program for computing the lucky numbers less than N is given in Fig. 6.5.

The sieve for generating lucky numbers is much less adaptable to execution by computer than a modular sieve since at each step the culls are determined relative to a *variable* set, the set L of Fig. 6.5, rather than to a *constant* set, the set $s_0 = \{0,1,...,K\}$ of Fig. 6.1. Sieving processes of the type exhibited in Fig. 6.5 have been used extensively by Wunderlich as part of his studies of sieve-generated sequences [195]. In order to shorten the time-consuming search operations of lucky-like sieves (lines L.3 and L.6 of Fig. 6.5), Wunderlich [196] suggests that, along with L in bit-pattern representation, a running tally of the number of elements in certain fixed portions of L be kept. For example, it is useful to maintain a list $T_{0 \text{ to } H}$, where $T_0 = 0$ and $T_h = |L \cap hK|$ for $h = 1$ to H. (H and K are parameters whose product is N.) On a computer with a w bit word, it is convenient to take $K = w$; then the specific counting of ones need only be undertaken through a single word—see § 2.2.3.) With such a model the operation "find $(L)_k$" becomes a two-step process: (1) find the portion of L in which the kth 1-bit of L resides (i.e. cal-

L.1 $L = (\text{set}) \; N$

L.2   for $i = 2, 3, \ldots,$

L.3 $j = (L)_i$

L.4 $c = \{ \; \}$

L.5   for $k = j, \ldots, k+j \to k, \ldots:$

L.6 $x = (L)_k$

L.7   if nonexistent: exit from loop

L.8 $c \cup \{x\} \to c$

L.9   if $c = \{ \; \}$: exit from loop

L.10 $L \sim c \to L$

L.11   "L is now the set of luckies $<N$."

FIG. 6.5. Lucky number generation.

culate h such that $T_{h-1} < k \leqslant T_h$), and (2) find the position of the $(k - T_{h-1})$th 1-bit in that portion—this position is $(L)_k$. This and other elaborate "tagged models" for sieving sets are discussed in Wunderlich's paper.

EXERCISES

1. Write a recursive version of the sieve of Eratosthenes. Use the indexed set of odd numbers as the universe.

2. Generalize the program of Fig. 6.4 so that success of the test at line G.5 initiates backtracking, so that failure to reach line G.8 would indicate that Goldbach's conjecture is false.

3. Write a program for generating lucky numbers which incorporates Wunderlich's idea of tagging. Do not forget to adjust the tags, the T_h for $h = 1$ to H, as numbers are eliminated from L.

*4. Find all numbers n, $n \leqslant 2 \times 10^6$, which cannot be written as the sum of four numbers of the form $\binom{i+2}{3}$. For a partial answer to this problem, see Salzer and Levine [183].

5. Write an efficient program to calculate the sequence $a_1, a_2, a_3, \ldots, a_n$, $n = 10^5$, where $a_1 = 1$, $a_2 = 2$, and a_i, $i > 2$, is the first integer larger than a_{i-1} which may be written as a sum $a_h + a_k$, $1 \leqslant h < k < i$, in only one way. (Compare exercise 7 of § 3.6.)

*6. Let S be a set consisting of the first t primes. Write a program to find all sets of triples $(n-1, n, n+1)$ of positive integers, with $n <$ given N, such that all of these integers belong to S. (See [175].)

7. As pointed out by Stein and Stein [188], there exists a simple algorithm for calculating the kth lucky number from the luckies $\leqslant k$. Devise an algorithm and write a program for this calculation.

8. Construct a sieve which produces irreducible polynomials modulo 2. Compare the efficiency of your algorithm with that of the program of Fig. 3.4.

6.2. Sorting Techniques

The clerical task of rearranging N objects so that N "keys" (one key is associated with each object) are in lexicographic order is known as *sorting*. Sorting was one of the first routine jobs considered for high-speed electronic computers (see [163]); it has been studied extensively since then (see [164], for instance). On a computer the keys are numbers, usually integers, given in standard hardware representation so that comparisons between them may be made easily. The objects are arbitrary, although frequently they are themselves numbers and serve as their own keys.

Sorting and combinatorial computing have a trinal affinity: (1) sorting is used as a tool in many combinatorial algorithms; (2) many interesting theoretical problems of a combinatorial nature arise in connection with sorting processes; (3) the programming of sorting procedures indeed typifies combinatorial programming. In this section we are concerned with the first of these relationships; we present programs to be used as procedures in applications given later. The second relationship is pursued in § 7.2. Regarding the third relationship, several important sorting schemes are left as exercises on which the reader may try his own combinatorial programming skills.

The literature on sorting is voluminous. This is so both because of the varied requirements of sorting applications—N is large or N is small, the given objects are partially sorted or in random order, etc.—and because various schemes are more suited to one computer system than to another. For our purposes, N is reasonably small, at most a few hundred, hence we are concerned with "internal" (as opposed to "tape") sorting. The procedures given here are quite fast and generally adequate for combinatorial computing on most machines. As always, however, when a long-running problem with extensive sorting arises, a little time spent in method-research usually pays for itself.

6.2.1. *Pigeonholing—Sort by Address Calculation*

Consider first the task of sorting N given keys $A_0, A_1, \ldots, A_{N-1}$ into increasing numerical order, forming the vector $(A_{p_0}, A_{p_1}, \ldots, A_{p_{N-1}})$, where $A_{p_0} \leqslant A_{p_1} \leqslant \ldots \leqslant A_{p_{N-1}}$. If the keys are distinct integers from "thin" universe (i.e. $A_{p_{N-1}} - A_{p_0}$ is not too large), then the possible values for the keys may be associated with consecutive words, or bits, of the computer called "pigeonholes". To sort a subset of keys, each key is placed in its assigned pigeonhole and then the pigeonholes are scanned for occupancy. Such a technique is, of course, implicit in our language for combinatorial computing, the vector $((S)_1, (S)_2, \ldots, (S)_N)$ being the sorted version of the set $S = \{A_0, A_1, \ldots, A_{N-1}\}$. The compact bit-pattern representation for sets greatly increases the applicability of this method. (Of course, if the set S is given as an ordered list (§ 3.2.2), there is not a specific slot assigned to each possible element, but the list may be kept sorted by use of the *adjoin*( ) procedure—see Fig. 3.3.)

A generalization of this method which may be applied when the keys are not distinct or when the universe is not thin is called sorting by address calculation (see, particularly,

A.0 $sort(N,\text{``}A\text{''};\text{``}B\text{''})$

A.1 $K = 2N$; for $i = 0$ to $K+1$: $B_i = 0$

A.2 $m = MIN_i(A_i)$; $M = MAX_i(A_i)$

A.3 $d = \dfrac{K-1}{M-m}$

A.4   for i:

A.5 $j = [d(A_i-m)]+1$

A.6   if $B_j = 0$: $B_j = A_i$; reiterate

A.7   if $A_i \geqslant B_j$:

A.8 $k^* = $ first $k = j+1, j+2, \ldots,$ such that $B_k = 0$ or $A_i \leqslant B_k$

A.9 $C = B_{k*}$; $B_{k*} = A_i$

A.10   for $k = k^*, k^*+1, \ldots,$ until $C = 0$: $C \leftrightarrow B_{k+1}$

A.11   if $k = K+1$:

A.12 $C = B_k$; $B_k = 0$

A.13   for $k^* = k, k-1, \ldots,$ until $C = 0$: $C \leftrightarrow B_{k*-1}$

A.14   otherwise:

A.15 $k^* = $ first $k = j-1, j-2, \ldots,$ such that $B_k = 0$ or $A_i \geqslant B_k$

A.16 $C = B_{k*}$; $B_{k*} = A_i$

A.17   for $k = k^*, k^*-1, \ldots,$ until $C = 0$: $C \leftrightarrow B_{k-1}$

A.18   if $k = 0$:

A.19 $C = B_0$; $B_0 = 0$

A.20   for $k^* = 0, 1, \ldots,$ until $C = 0$: $C \leftrightarrow B_{k*+1}$

A.21 $i = 0$

A.22   for $j = 1$ to K such that $B_j \neq 0$:

A.23 $B_i = B_j$; $i+1 \rightarrow i$

A.24   exit from procedure

FIG. 6.6. Sort by address calculation.

Isaac and Singleton [170] and Flores [159]). In this method, the assignment of possible key values to (word) pigeonholes is calculated from a (usually linear) formula. This assignment may not be unique, so the allocation of keys to pigeonholes is more involved. When, during this allocation, it occurs that the pigeonhole associated with a key is already occupied, it is necessary to find this key's correct relative position and possibly shift some nearby keys so that the new key may be inserted. The correct relative position of the sorted keys is maintained at each step, so, as before, when the allocation is complete, a scan of the pigeonholes produces a sorted list of keys.

A straightforward procedure which sorts by address calculation is exhibited in Fig. 6.6. The unsorted keys are given in the array $A_{1 \text{ to } N}$. While a few as N slots are needed into which the keys are allocated, much less shifting is required if, in fact, more than N are made available. Flores suggests using about $2N$. Line A.1 prepares the array B for this purpose. (We use the value zero to denote a vacant slot; if a key may equal zero, a different label must be used for this purpose.) The formulae at line A.2 compute the range of the keys for use in calculating the approximate pigeonhole of key A_i by the simple linear interpolation formula at lines A.3 and A.5. The loop beginning at line A.4 distributes the keys. For most keys, the test at line A.6 is successful and no moving of keys is required. If, however, the slot associated with A_i is occupied, the correct relative position of A_i must be found; this is accomplished by lines A.8 and A.9 when $A_i \geqslant B_j$ (or lines A.15 and A.16 when $A_i < B_j$). Then a vacant slot for A_i is forced open by shifting the "succeeding" previously sorted keys into a nearby gap; this is done at lines A.10 (or A.17). To simplify the logic of this adjustment, vacant slots are maintained at B_0 and B_{K+1} by the tests and program beginning at line A.11 and line A.18. When all keys have been allocated, the gaps in B are removed by the iteration of lines A.21 through A.23 leaving sorted keys at $B_0, B_1, \ldots, B_{N-1}$.

Sorting by address calculation is usually extremely fast, especially so when the distribution of keys is essentially uniform. Its chief disadvantage is the large amount of auxiliary storage, the B array, which is needed.

6.2.2. Merge–Exchange Sorting—Shell's Method

By performing a series of exchanges it is possible to sort N keys $A_{1 \text{ to } N}$ *without* using extra storage. First, A_2 is compared with A_1 and, if necessary, exchanged with A_1. Then A_3 is compared with (the possibly new) A_2 and perhaps exchanged with it. If an exchange was necessary, then a further comparison, and possible exchange, with A_1 is performed. In a similar manner A_4 is placed in its corect position relative to $A_1 \leqslant A_2 \leqslant A_3$, and so forth; the list is sorted when A_N has been correctly placed.

The number of comparisons and exchanges required to find the relative place for a key, by this "sinking" technique, is significantly smaller when the keys are initially almost in order. D. L. Shell suggests a method [184] in which the sinking technique is recursively applied to longer and longer subsequences of the main sequence. First, the sequences $(A_1, A_{1+d}), (A_2, A_{2+d}), \ldots, (A_d, A_{2d})$, where d is about $N/2$, are sorted. Then the sequences $(A_1, A_{1+d}, A_{1+2d}, A_{1+3d}), (A_2, A_{2+d}, A_{2+2d}, A_{2+3d}), \ldots$, where d is about $N/4$, are sorted. The process is repeated for approximately $\log_2 N$ steps; the last step (when $d = 1$) sorts the entire sequence.

A program for Shell's method is given in Fig. 6.7. The quantifier at line S.2 controls the basic recursion, the procedure *distance* (d) producing the appropriate value for d at each step. Here we have used a formula suggested by Frank and Lazarus [160] which makes

S.0   *sort*$(N;$ "A")

S.1 $d = N$

S.2   for each *distance*$(d) \rightarrow d$:

S.3   for $i = 1$ to d:

S.4   for $j = i, i+d, \ldots$, until $j+d > N$:

S.5   for $k = j, j-d, \ldots, i$, while $A_k > A_{k+d}$:

S.6 $A_{k+d} \leftrightarrow A_k$

S.7   "At this point $A_{1 \text{ to } N}$ is sorted."

S.8   exit from procedure

S.D.0   *distance*(d)

S.D.1   if $d = 1$: *Exit* $= 1$; exit from procedure

S.D.2   *Exit* $= 0$; *distance*$(\ ) = 2\left[\dfrac{d}{4}\right] + 1$

S.D.3   exit from procedure

FIG. 6.7. The Frank–Lazarus variation of Shell's sorting algorithm.

d odd, thereby preventing subsequences from being sorted independently prior to being merged. Line S.3 controls the iteration over the d subsequences at each step. Lines S.4, S.5, and S.6 sort a particular subsequence—line S.4 specifies a key while the iteration and exchange of lines S.5 and S.6 locate its correct relative position.

Shell's method is commendable for its simplicity as well as for its economy of storage. Its speed also compares favorably with that of other techniques of its kind (see Hibbard [168]). Of course, if used extensively, special care should be taken to tailor the "inner loop" (lines S.5 and S.6 of Fig. 6.7) for a particular computer system.

6.2.3. *The Sort Permutation*

So far we have only considered the task of sorting N numerical keys. While this represents the important case for combinatorial computing, one is occasionally confronted with the job of rearranging more complex objects according to some given order relation. Because of the awkwardness and inefficiency involved in moving large objects about within the memory of a computer, it is usually best first to calculate a "sort permutation" which describes the correct order of the objects and then to rearrange the objects as directed by this permutation.

Let $C_0, C_1, \ldots, C_{N-1}$ be N "comparable" configurations; that is, there exists an order relation expressed by the symbols $=, <, >, \leqslant$, etc., such that for $i \in N$ and $j \in N$ either $C_i = C_j$, $C_i < C_j$ or $C_i > C_j$. A *sort permutation* is a permutation $p_{0 \text{ to } N-1}$ of $(0,1,\ldots,N-1)$ such that $C_{p_0} \leqslant C_{p_1} \leqslant \ldots \leqslant C_{p_{N-1}}$. Note that if configurations are the numbers $0, 1, \ldots, N-1$ with the natural order relation, and are distinct, then p is merely the permutation inverse to C.

Such a permutation p may be constructed by performing upon the sequence $0, 1, \ldots, N-1$ the transmissions required to sort the configurations. For instance, Shell's method of Fig. 6.7 may be used to construct a sort permutation $A_{1 \text{ to } N}(\equiv p_{0 \text{ to } N-1})$ provided the comparison of line S.5 is replaced by a comparison among the C's:

$$\text{while } C_{A_k} > C_{A_{k+d}}$$

for example. Alternatively, if the configurations have *simple* keys $A_{1 \text{ to } N}$ which determine their proper order, then array $(p_0, p_1, \ldots, p_{N-1})$, initialized to $(0, 1, \ldots, N-1)$ can be re-arranged exactly as are the keys. For example, the statement "$p_{k+d-1} \leftrightarrow p_{k-1}$" can be added to line S.6. Possibly these marks can even be initially combined with the keys (in the sense of a fractional word representation) in a portion of the word which does not significantly affect the comparisons. They are then carried along by the transmissions of line S.6 and later extracted as the sort permutation.

P.1 $S = (\text{set}) N$

P.2   for $i \in S$:

P.3 $C_{p_i} \to D$

P.4   for $k = p_i, \ldots, p_k \to k, \ldots,$ until $k = i$:

P.5 $S \sim \{k\} \to S$

P.6 $C_{p_k} \to C_k$

P.7 $D \to C_i$

FIG. 6.8. Sorting via a precalculated permutation.

The rearrangement of the configurations as prescribed by a precomputed sort permutation $p_{0 \text{ to } N-1}$ is accomplished by a program with the structure exhibited in Fig. 6.8. As an example, let $C_0 = c$, $C_1 = d$, $C_2 = a$, $C_3 = e$, and $C_4 = b$, where $a < b < c < d < e$. The sort permutation is $(2,4,0,1,3)$ which has the cycles $(0\ 2)$ and $(1\ 4\ 3)$. The program of Fig. 6.8 interchanges C_0 and C_2 by means of the transmissions $C_2 \to D$, $C_0 \to C_2$, and $D \to C_0$. The second cycle is accomplished with the transmissions $C_1 \to D$, $C_4 \to C_1$, $C_3 \to C_4$, and $D \to C_3$. In general a cycle $(h_1 h_2 \ldots h_m)$, which says C_{h_2} has final position h_1, C_{h_3} has final position h_2, $\ldots$, C_{h_1} has final position h_m, is accomplished with $m+1$ transmissions. These transmissions are indicated symbolically in Fig. 6.8 by the simple assignment statements at lines P.3, P.6, and P.7; in practice, of course, such transmissions are more involved.

Since the program of Fig. 6.8 is independent of the origin of the permutation $p_{0 \text{ to } N-1}$, it has applicability independent of sorting. This program structure has already been seen in the program of Fig. 3.5 for computing the cycles of a permutation. Another application occurs with respect to matrix transposition discussed in § 6.5.

EXERCISES

1. The program of Fig. 6.9 merges two sorted lists $A_{1 \text{ to } n}$ and $B_{1 \text{ to } n}$ of positive integers (in decreasing order) into the sorted list $C_{1 \text{ to } 2n}$. Write a program which applies this technique to successively larger sublists of a given unsorted array, eventually producing a single sorted list. This method is known as *two-way merge sorting*.

$$a = A_1 \; ; \; b = B_1 \; ; \; t = 1; j = 1$$

for $k = 1$ to $2n$:

    if $a > b$:

        $C_k = a; \; i{+}1 \rightarrow i$

        if $i \leqslant n$: $a = A_i$; otherwise: $a = 0$

    otherwise:

        $C_k = b; \; j{+}1 \rightarrow j$

        if $j \leqslant n$: $b = B_j$; otherwise: $b = 0$

FIG. 6.9. The merging of two sorted lists.

*2. Two-way merge sorting requires about $N \log_2 N$ object transmissions since if the sublists double in size at each step, there are about $\log_2 N$ steps and each object is moved from one of two sublists onto the larger sublist at each step.

    (a) For 2^K unsorted keys devise a two-way merge procedure which makes a precise record of the $2^K K$ transmissions required to sort the keys. Note that this record may be kept as a mask of $2^K K$ bits since a transmission consists of moving a key from one of two lists.

    (b) Write a program which uses the mask produced by the above procedure to sort 2^K objects associated with the keys.

*3. Sorting whereby the keys are classified and distributed first by their most significant digit, then within each resulting class by their second digit, and so forth, is known as *radix sorting*.

    (a) Write a program, using as much auxiliary storage as desired, which sorts N not necessarily distinct natural numbers $A_1, A_2, \ldots, A_N$, each less than 2^K, by a base 2 radix sorting scheme.

    (b) Radix sorting schemes in which the movement of the keys is accomplished with interchanges (e.g. $A_i \leftrightarrow A_j$), hence which require minimal auxiliary storage, are known as *radix-exchange* methods. Write a base 2 radix-exchange program for sorting N keys as described in part (a).

4. A useful *indirect* means of calculating serial numbers for complex objects consists of the application of binary search (§ 3.2.2) to a presorted list of the objects. Adapt this approach to a serial number calculation scheme for partitions of a set. Have your scheme produce serial numbers different from those given by (5.13).

*5. Note that the iteration quantifier at line S.5 of Fig. 6.7 can be eliminated when the *distance( )* algorithm halves d at each step. Compare, empirically, this approach (Shell's original algorithm) with the algorithm given in the text.

6.3. Procedures Concerned with Connectedness

The concept of connectedness is encountered in a multitude of topological and puzzle-like problems. "What is the cyclic structure of this permutation?" or more generally "What are the connected components of this graph?" are typical questions asked in combinatorial programming. While it happens that general-purpose digital computers are not very well suited for connectivity determination, there are a few algorithmic tricks available which improve the situation somewhat. (The question of connectedness deserves attention by computer hardware designers. Many natural methods of combinatorial computing are rendered ineffectual by the inefficiency of connectivity procedures.)

6.3.1. *The Connectivity Matrix—Warshall's Method*

Vertex u of a graph is *connected* to vertex v of the graph if there exists a path from u to v (§ 3.4.4). The essential feature of this concept of connectedness is the *transitivity* of the binary relation defined on the set of vertices; that is, if u is connected to v and v is connected to w, then u is connected to w. In many cases, the connective relation is also *reflexive* (i.e. u is connected to u) and *symmetric* (i.e. u is connected to v implies v is connected to u). This need not be true in general, however; for example, "unilateral" connectedness of a directed graph is neither reflexive nor symmetric.

To an arbitrary graph G represented by the adjacency matrix $G_{0 \text{ to } n-1}$ one may associate a unique graph G^* represented by $G^*_{0 \text{ to } n-1}$ where $i \in G^*_j$ if and only if there is a path in G from j to i. The incidence matrix G^* is said to be the *connectivity* (or *reachability*) matrix associated with G. An important manipulation is to calculate the connectivity matrix for a given incidence matrix.

"Warshall's method"

for $i = 0$ to $n-1$:

for $j = 0$ to $n-1$ such that $i \in A_j$:

$A_j \cup A_i \rightarrow A_j$

"At this point A is transitive"

FIG. 6.10. A program for Warshall's component generation scheme.

The best existing method for this construction is due to Warshall [190]. Let the incidence matrix A be given as a 1-array of sets $A_{0 \text{ to } n-1}$ such that $A_j, j \in n$, is the set of elements which j precedes, or, alternatively, the set of elements to which j is connected by a path of length one. Warshall's method, given in Fig. 6.10, transforms A into its own connectivity matrix by a series of union operations. In words, the algorithm says to adjoin A_i (i.e. those elements which i so far precedes) to A_j for each j so far preceding i, for each i in turn.

Consider the example of Fig. 6.11. As indicated in the matrix by X's, we are given initially that 0 precedes 3, 1 precedes 4, 2 precedes 0, 3 precedes 1, and 5 precedes 2. As the algorithm progresses, the higher order connections are discovered; the value of i for the step at which a particular connection is discovered is given in the appropriate matrix slot. Consider the step for $i = 2$. At this point, 5 is known to precede 2 and since 2 precedes 0, we deduce that 5 precedes 0, hence also 3. The other deductions about 5

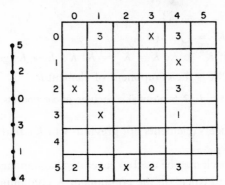

FIG. 6.11. Understanding Warshall's method.

(and about 2 and 0) come later, namely when $i = 3$. In general, for each value of i, all connections are established for the path

where h is the nearest preceding element not yet considered (i.e. $h > i$ and for all $x \in A_h$ such that $i \in A_x, x > i$) and k is the nearest succeeding element not yet considered (i.e. $k < i$ and for all $y \in A_i$ such that $k \in A_y, y < i$). Such an analysis applies to each path of the relation. Therefore, when all elements have been considered, all possible connections are established and the matrix is transitive. (A formal proof of this result appears in the cited paper [190].)

6.3.2. *Sequential Generation of Components*

Now let the 1-array of sets $A_{0 \text{ to } n-1}$ represent a *symmetric* binary relation on the set $S = \{0, 1, \ldots, n-1\}$, i.e. A is the adjacency matrix for an *undirected* graph with n vertices. A *component* of S is a subset T such that $i \in T$ and $j \in T$ implies i is connected to j; moreover, $i \in T$, $k \in S$, and i connected to k implies that k is in T. (With a little care useful analogous definitions can be made for directed graphs.) It is usually convenient to consider an isolated vertex as a component, although by strict definition it is so only if a loop appears on it. Note that by transitivity a vertex is always connected to itself if its component contains other vertices.

If C is the connectivity matrix associated with A, then the component containing vertex k is simply C_k. Since Warshall's algorithm gives the connectivity matrix only when completed, its use may be inefficient when only a particular component is needed or when, due to lack of storage space, the components must be generated one at a time. An alternative method in which only the component containing element k is formed is exhibited in Fig. 6.12. The set K is ultimately the component containing k; the set $\bar{K}$ consists of recently discovered elements of the component; the set K^*, formed at line C.2, consists of all elements connected to any element of $\bar{K}$. The process continues as long

C.1 $K, \bar{K} = \{k\}$

C.2   #2 $K^* = \bigcup_{i \in \bar{K}}(A_i)$

C.3   if $K^* \not\subset K$: $\bar{K} = K^* \sim K$; $K \cup \bar{K} \to K$; go to #2

         "K is the component of element k."

FIG. 6.12. Sequential generation of components.

as K^* contains new elements (elements not previously in K), that is, as long as the test of line C.3 is successful.

Of course, this approach may be used to generate all components of the set S. We may initially set $k = (S)_1$, and calculate the component K containing the element k by the program of Fig. 6.12, then let $S \sim K \to S$, again set $k = (S)_1$ and calculate a component, this process being repeated until S is empty. Such an approach is used in Fig. 7.11 (specifically, by the *component*( ) procedure) for calculating the "weak" components of a partial ordering.

We have so far tacitly assumed that a relation is given by an incidence matrix. There are situations in which it is *in*convenient to construct an incidence model for a binary relation. Such situations often arise with respect to geometric connectedness—for instance, "rookwise" connectedness of squares on a chessboard. Although the exact programs of Figs. 6.11 and 6.12 cannot be used in these cases, their structure can still be adopted. An example of this is given in Fig. 6.13. The generation procedure *polyomino*( ) forms

P.0   *polyomino*(n)

P.1   for $j = 0$ to $n+1$: $K_j = \{\ \}$

P.2 $x_0 = $ first $i = 0$ to $n+1 \ni |B_i| \neq 0$

P.3   if nonexistent: *Exit* = 1; exit from procedure

P.4 $y_0 = (B_{x_0})_1$

P.5 $H = 1$

P.6   for $h = 0$ to $H-1$:

P.7   execute *test*(x_h, y_h+1)

P.8   execute *test*(x_h, y_h-1)

P.9   execute *test*(x_h+1, y_h)

P.10   execute *test*(x_h-1, y_h)

P.11   *Exit* = 0; *polyomino*( ) = K; exit from procedure

T.0   *test*(i, j)

T.1   if $j \in B_i$:

T.2 $x_H = i$; $y_H = j$

T.3 $K_i \cup \{j\} \to K_i$; $B_i \sim \{j\} \to B_i$

T.4 $H+1 \to H$

T.5   exit from procedure

FIG. 6.13. Application of component generation to a "geometric" zero–one matrix.

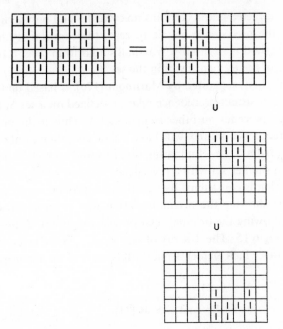

FIG. 6.14. The component decomposition of a zero–one matrix.

in turn each rookwise connected component of a shape defined by the 1-bits in a zero–one matrix $B_{1 \text{ to } n}$ (actually $B_{0 \text{ to } n+1}$, since a border of 0-bits is used—see § 3.6.2). The approach is roughly that of Fig. 6.12. The decomposition (into three components) of a particular shape is illustrated in Fig. 6.14. Note that this program destroys the original shape as components are formed (line T.3).

6.3.3. *Consistency Determination (Cycle Detection)*

A binary relation which is "anti-symmetric" (i.e. *a* is related to *b* implies *b* is not related to *a*, when *a* and *b* are distinct) is called a *precedence relation*; the corresponding incidence matrix is called a *precedence matrix*. In a precedence matrix, $M_{i,j} = 1$ implies $M_{j,i} = 0$ for $i \neq j$. A precedence relation (and matrix) is *consistent* if its connectivity matrix is also a precedence matrix. A consistent precedence relation on a set S defines a *partial ordering* of the elements of S; all order relations are given by non-zero entries of the connectivity matrix. In most mathematical expositions a partial ordering is also considered to be reflexive. In our work reflexivity is of little importance and most relevant algorithms simply ignore the entries along the main diagonal. A set with a reflexive, anti-symmetric, transitive binary relation defined on it is called a *poset*.

A poset is, of course, a directed graph without cycles; consistency determination is thus equivalent to the detection of no cycles in an arbitrary directed graph. This task arises in various contexts. From the above discussion we note that a precedence matrix is consistent (i.e. a directed graph contains no cycles) if and only if, having initialized the main diagonal of the precedence matrix to zero, the connectivity matrix retains those zeros. Thus one method is to modify Warshall's algorithm (Fig. 6.10) so that it detects on the appearance of the element j within the set A_j.

An alternate method, suggested by Marimont [176], is based on the (nearly obvious) fact that for every consistent precedence relation defined on a set S, there is at least one element of S which precedes no other elements of S. Thus if the precedence relation is consistent, the precedence matrix will have a row containing only zeros. Deleting the associated element from S (i.e. eliminating that row and corresponding column from the matrix) produces a smaller set with a precedence relation which is consistent if and only if the original relation is consistent. By iterating this process we either produce a matrix with no zero rows, showing the original relation was inconsistent, or we eliminate all elements from S, showing the original relation was consistent. A program for this algorithm is shown in Fig 6.15. The 1-array of sets $A_{0 \text{ to } n-1}$ is the given precedence relation. The mask S indicates which elements may still be part of an inconsistency.

$$S = (\text{set})\, n$$

$$\text{while } \exists_{i \in S}(A_i \cap S = \{\ \}):$$

$$S \sim \{i\} \to S$$

$$\text{if } S = \{\ \}: \text{PROGRAM “A is consistent”}$$

$$\text{otherwise: PROGRAM “A is inconsistent”}$$

Fig. 6.15. A program for the determination of poset consistency.

Exercises

1. Write a program which generates connected components of a linear graph directly from an edge-vertex incidence matrix.

2. From a given matrix $A1_{0 \text{ to } n-1}$, define $A2_{0 \text{ to } n-1}$ by $j \in A2_i$ if and only if $\exists_{k \in n} (j \in A1_k$ and $k \in A1_i)$ is true, i.e. if and only if there is a path of length two from i to j in $A1$. The "powers" $A3$, $A4$, ... can be defined similarly. Eventually, since the graph is finite, the powers converge to the reachability matrix. Write an efficient program to calculate

$$MAX_{i;j}(MIN_{\text{ALL PATHS FROM } i \text{ to } j}(\text{LENGTH OF PATH}))$$

for a given square incidence matrix.

3. Let $A1_{0 \text{ to } n-1}$ be an arbitrary (not necessarily square) incidence matrix. The "first derived" incidence matrix of $A1$, call it $A2$, may be defined by

$$\text{for } i \in n: A2_i = \bigcup_{k \in n \ni A1_i \cap A1_k \neq \{\ \}} (A1_k).$$

Similarly, higher order connectivity relations, and in fact a *connectivity matrix*, can be defined relative to an arbitrary incidence system. Discuss the procedures presented in this section as they might apply to general incidence matrices. (See introduction to §6.4.)

4. Give reasonable definitions for the *strong component* and *weak component* of a vertex v in a partly directed graph.

5. Describe the connectivity matrices for the following configurations:

(a) A connected undirected graph with n vertices.

(b) A directed graph in which each vertex is part of a cycle.

(c) A poset in which all elements are comparable, i.e. either $a < b$ or $b < a$ for all $a \neq b$ in the set (such a "totally ordered" set is called a *chain*).

6. Write a program to generate rookwise connected components of a given zero–one matrix $B_{0 \text{ to } n-1}$ by applying the program of Fig. 6.12 to the appropriate graph-theoretic model (see § 3.6).

7. Write a program for forming rookwise connected components of a zero–one matrix using Warshall's approach directly, i.e. without first constructing a graph-theoretic model.

8. Program Marimont's method of testing for consistency by using the existence of a maximal as well as a minimal element in a poset.

*9. Write a program to generate all k vertex connected subgraphs which contain a specified vertex of a given graph.

10. Write a program to generate all maximal cliques with k vertices in a given graph $G_{0 \text{ to } n-1}$.

*11. A *spanning tree* T of a connected undirected graph G is a tree subgraph which contains all vertices of G (T and G are edge sets). If $e \in G \sim T$, then the subgraph $T \cup \{e\}$ contains a circuit (the reader should verify this for himself). The set of circuits obtained by adjoining in turn each edge in $G \sim T$ is called a *fundamental* (or *basic*) *set of cycles*. This terminology is used since every circuit of G is a symmetric difference of circuits from this basic set (see [215], for instance).

(a) Write a program to generate a spanning tree and (simultaneously) a basic cycle set for a given graph.

(b) Write a program to generate all circuits in the graph.

6.4. Finite Set Covering

We now consider a class of abstract problems which model a large number of combinatorial situations. Let $S_{0 \text{ to } m-1}$ represent an incidence system (§ 3.4.3) in which the objects are the numbers $0, 1, \ldots, n-1$. A *cover* C of S is a subset of $\{S_0, S_1, \ldots, S_{m-1}\}$, the union of whose elements is $\{0, 1, \ldots, n-1\}$, i.e.

$$C \subset \{S_0, S_1, \ldots, S_{m-1}\}$$

and

$$\bigcup_{c \in C} (c) = \{0, 1, \ldots, n-1\}.$$

Although a slight abuse of language, it is sometimes convenient to view the subset of $\{0, 1, \ldots, m-1\}$ which describes the cover as the cover itself; that is, I is called a cover when

$$I \subset (\text{set}) \, m,$$

and

$$\bigcup_{i \in I} (S_i) = (\text{set}) \, n \tag{6.3}$$

are true. We are concerned here with the structure of algorithms which generate and/or enumerate covers for a given incidence system subject to various restrictions on their constitution. The most common restrictions are "disjointness" (§ 6.4.2) and "irredundance" (§ 6.4.3). The construction of "minimal" covers is pursued further in § 7.5.

These covering problems are closely related to problems from "matching theory", a branch of combinatorics. See Liu [28] (and exercise 9) for an introduction to this important subject.

6.4.1. *The Basic Branching Algorithm*

A naïve method of generating covers is to generate all subsets of $\{0,1,\ldots,m-1\}$, eliminating in turn each one that does not satisfy condition (6.3). As discussed in Chapter 4, however, it is much better to merge the test with the generation itself. A basic backtrack algorithm for cover generation appears in Fig. 6.16. Actually this algorithm solves a slightly more general problem: For N a given subset of $\{0,1,\ldots,n-1\}$, generate all sets C^* which satisfy

$$C^* \subset \{0,1,\ldots,m-1\}$$

and

$$N \subset \bigcup_{i \in C^*}(S_i) \subset \{0,1,\ldots,n-1\}. \tag{6.4}$$

The idea of this algorithm is to generate certain "basic" covers with a nest of iterations, each of these basic covers giving rise to several "complete" covers via the iteration at the center of the nest (line C.11). The nest is a systematic selection of blocks whose union covers all objects in N.

C.1 $G_0 = (\text{set})\ m;\ H_0 = \{\ \}$

C.2   with $k = 0, 1, \ldots$, as $C = \{\ \}, \{\ \}\cup\{b_{k-1}\}, \ldots$, until $H_k \supset N$:

C.3 $\cdot\cdot$

C.4 $j = (N \sim H_k)_1$

C.5 $B_k = V_j \cap G_k$

C.6   for $b_k \in B_k$:

C.7 $H_{k+1} = H_k \cup S_{b_k}$

C.8 $G_k \sim \{b_k\} \to G_k$

C.9 $G_{k+1} = G_k$

C.10 $\cdot\cdot$

C.11   for $D \subset G_k$: USE $C^* = D \cup C$ AS A COVER

FIG. 6.16. The basic cover generation program.

Notationally, S is the given incidence matrix and V is its transpose, i.e. for $j \in N$: $V_j = \bigcup_{i \in m\ \ni\ j \in S_i}(\{i\})$. The set C becomes a basic cover. At each step (i.e. for each k), j is the smallest as yet uncovered object and B_k is the set of blocks available for covering that object; b_k is the block actually used to do the covering. The set H_k gives the objects covered by $b_0, b_1, \ldots, b_{k-1}$ while G_k gives those blocks still available for use in the cover at later steps. Note that b_k is deleted from G_k itself (line C.8), as well as from G_{k+1} (line C.9); this insures that duplicate covers are not generated. At the center of the nest, each subset D of those blocks which are still available is adjoined to the basic cover to form a complete cover. If covers are merely being counted and not explicitly generated, then the iteration at line C.11 would be replaced by an addition of $2^{|G_k|}$ to the tally.

Structurally, this algorithm is similar to the branching process used by McCluskey [177] and Roth [182] to simplify Boolean expressions (see § 7.5). It is also closely related to the branch-and-bound method discussed in § 4.3. The branching process recursively resolves the generation problem into a pair of simpler problems: the generation of covers which contain a particular block and the generation of covers which do not contain that block.

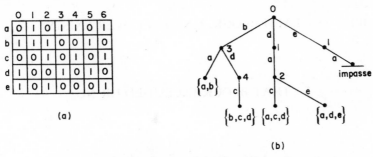

(a)

(b)

FIG. 6.17. Example of search tree for basic cover generation.

We illustrate this point of view with the example of Fig. 6.17. A 5×7 incidence matrix is given in (a) with the blocks labeled a, b, c, d, e instead of 0, 1, 2, 3, 4 to avoid confusion with the numerical labels of the objects. The search tree associated with the cover generation appears in (b). Nodes of the tree correspond to uncovered objects while links emanating from a node correspond to available blocks which contain that object. In the example, 0 is the first uncovered object: it is covered by blocks b, d and e. Link labeled b terminates at node labeled 3 since 3 is the first uncovered object when block b is used to cover object 0. Using a to cover 3, we then have a basic cover $\{a,b\}$. We generate complete covers by adjoining in turn $\{\ \}, \{c\}, \{d\}$, and $\{c,d\}$. As we backtrack, and move from link a to link d for covering 3, we are "branching" on a, i.e. we will now generate covers which do *not* contain a (but still contain b). Thus a is not available for covering vertex 4. Once the other branch from node 0 is taken, a again becomes available since we are then assuming that b is not in the cover. In general, backtracking—i.e. passing from one link to the next link—corresponds to branching on the block of the previous link. In Fig. 6.16 the sets G_k control this branching process; elements still in G_k at line C.5 are those blocks which have not yet been branched upon.

Since the incidence matrix formulation of Fig. 6.16 is sometimes inappropriate, it is important to understand the basic structure of cover generation algorithms. The two program structures given in Fig. 6.18 illustrate the two main viewpoints for cover generation. In segment A, vertical movement in the tree (nest) amounts to finding the next uncovered object, while horizontal movement (the general iteration of the nest) corresponds to finding ways in which that object may be covered. In segment B, each level of

A.1   with EACH (NEXT) UNCOVERED OBJECT:

    •

    •

    •

for EACH BLOCK COVERING THAT OBJECT:

      •

USE COVER

B.1   with EACH BLOCK, UNTIL A COVER EXISTS:

   •   •

   •   •

   •   •

for THAT BLOCK INCLUDED OR NOT:

      •

      •

USE COVER

FIG. 6.18. The two fundamental viewpoints for cover generation algorithms.

the *binary* tree is associated with a particular block, the two links emanating from a node corresponding to the inclusion or exclusion of the associated block in the partial cover. Restrictions on the covers, or impasse detection in general, may be incorporated either in the nest generation procedure as illustrated by Fig. 6.19 of the next section or within the nest itself as illustrated by Fig. 6.16 here and Fig. 7.26 of § 7.5. Also, the reader should compare the structure of the TSP algorithm of Fig. 4.16 with segment B of Fig. 6.18.

6.4.2. *Disjoint Covers—Steiner Triple Systems*

Many problems require that the blocks in a cover be pairwise disjoint, i.e. that each object of N be contained in precisely one block of the cover. This restriction may be incorporated into the basic algorithm of Fig. 6.16 by replacing line C.9 by

$$G_{k+1} = G_k \sim Z_{b_k}$$

where

$$Z_{b_k} = \bigcup_{h \in S_{b_k}} (V_h)$$

is the precomputed set of blocks which intersect the block currently being adjoined to the cover. Also, the iteration at line C.11 should be deleted since only basic covers can be disjoint (when $N = $ (set) n, that is).

There are, of course, far fewer disjoint covers than unrestricted covers. In fact, it is possible to waste considerable time on branches of the search tree which contain no covers. Thus it generally is worth while to incorporate certain impasse detections into the basic program. One useful impasse detection is the test for uncovered objects which are not contained in any available block. This could be implemented by inserting the statement

$$\text{if } \exists_{h \in N} \sim H_k\left(V_h \cap G_k = \{ \ \}\right): \text{ exit from loop} \tag{6.5}$$

as the final statement of the body of the general loop, i.e. as line C.12 at the same indention level as line C.9. (It is instructive to compare this impasse detection with that suggested in exercise 2 of § 5.2; in fact, it is instructive to view restricted position permutation generation as a covering problem.)

As an example of disjoint covering, we consider the problem of constructing *Steiner number triple systems*, sets of $n(n-1)/6$ unordered triples of the labels 0, 1, ..., $n-1$ such that each pair of labels appears in exactly one triple. Such systems exist for a given n when and only when n is odd and $I = n(n-1)/6$ is an integer. The set $\{\{0,1,2\},\{0,3,4\}, \{0,5,6\},\{0,7,8\},\{1,3,5\},\{1,4,7\},\{1,6,8\},\{2,3,8\},\{2,4,6\},\{2,5,7\},\{3,6,7\},\{4,5,8\}\}$ is a system for $n = 9$. The enumeration of systems inequivalent under arbitrary permutation of the labels is an interesting problem which has been pursued with the help of computers [167].

By equating triples with blocks and pairs with objects, we have an incidence system in which an object is contained in a block if and only if the corresponding pair is a subset of the triple. A Steiner triple system is then a disjoint cover for this incidence system. We present in Fig. 6.19 a basic program for finding such a cover for arbitrary odd n (such that $n(n-1)/6$ is integral, of course). This program has the structure illustrated in segment A of Fig. 6.18 but does not use the incidence matrix formulation because that matrix is rather large and sparse.

A triple of the system is given by $T_i = \{a_i, b_i, c_i\}$, $i = 0$ to $I-1$. The sets M_j, $j = 0$ to $n-1$, give the labels with which label j has yet to be paired. The nest generation procedure *uncovered( )*—see § 1.5.2 and Fig. 1.14 for a discussion of the implementation of such a procedure—produces the next uncovered pair (lines U.7 and U.8) and a set of labels, C_{i+1}, determining the triples which cover that pair (line U.9). The disjointness of the cover follows from the use of M at lines U.8 and U.9. A legal system of triples has been constructed when the center of the nest is reached.

6.4.3. *Irredundant Covers*

For an arbitrary incidence matrix, the majority of covers produced by the basic procedure of Fig. 6.16 contain superfluous blocks, blocks which may be deleted from the cover without leaving any of the objects uncovered; such covers are *redundant*. As for disjoint covers, we may modify our basic procedure so that it produces only *irredundant* covers.

T.0 $steiner(n)$

T.1   for $j = 0$ to $n-1$: $M_j = n \sim \{j\}$

T.2 $I = n(n-1)/6$

T.3   with each $uncovered(\ )$:

T.4 $\cdot\ \cdot$

T.5   for $c_i \in C_i$:

T.6 $T_i = \{a_i,b_i,c_i\}$; $M_{a_i} \sim \{b_i,c_i\} \rightarrow M_{a_i}$

T.7 $M_{b_i} \sim \{a_i,c_i\} \rightarrow M_{b_i}$; $M_{c_i} \sim \{a_i,b_i\} \rightarrow M_{c_i}$

T.8 $\cdot\ \cdot$

T.9 $T_{0 \text{ to } I-1}$ IS THE TRIPLE SYSTEM

T.10 $M_{a_i} \cup \{b_i,c_i\} \rightarrow M_{a_i}$; $M_{b_i} \cup \{a_i,c_i\} \rightarrow M_{b_i}$

T.11 $M_{c_i} \cup \{a_i,b_i\} \rightarrow M_{c_i}$

U.0 $uncovered(\ )$

U.1   if $Entry1 = 1$ and $Entry2 = 0$;

U.2 $a_0 = 0$; $b_0 = 1$; $C_0 = M_0 \cap M_1$

U.3 $Exit = 0$; $i = 0$; exit from procedure

U.4   if $Entry2 = 0$:

U.5   if $i = I$: $Exit = 1$

U.6   otherwise: $Exit = 0$

U.7 $a_{i+1} = (\text{first } x = a_i, a_i+1, \ldots \ni M_x \neq \{\ \})$

U.8 $b_{i+1} = (M_{a_{i+1}})_1$

U.9 $C_{i+1} = M_{a_{i+1}} \cap M_{b_{i+1}}$

U.10 $i+1 \rightarrow i$

U.11   otherwise:

U.12   if $i = 0$: $Exit = 1$

U.13   otherwise: $Exit = 0$; $i-1 \rightarrow i$

U.14   exit from procedure

FIG. 6.19. Steiner number triple system generation.

First of all, the iteration over subsets of G_k at the center of the nest may be omitted, since C is already a cover. A basic cover, though likely to be irredundant, may still be redundant. Therefore the program must contain a test for redundancy. Rather than test each basic cover for redundancy, it is more efficient to eliminate redundancies as they occur during the construction of a cover. In this way, we prune from the tree large branches which can only produce redundant covers.

In the basic procedure, a block b_k is considered for membership in the subset C, which is to become a cover, only if it covers a previously uncovered object. When adjoined to C, however, it may render that subset redundant by also covering an object which a previously adjoined block was expressly chosen to cover. Therefore the redundancy test consists of asking if a previously adjoined block is now extraneous. If so, then we need not adjoin this new block, nor follow its derived branches, but may immediately try the next block which covers the uncovered object. This test may be accomplished by inserting the statements

$$\text{if } \exists_{x=k-1,\, k-2,\, \ldots,\, 0,\, \text{as } h=S_{b_k},\, S_{b_k}\cup S_{b_{x+1}},\, \ldots}\; [(H_x \cup h)\cap N = (H_{k+1}\cap N)]:$$

      reiterate

between lines C.8 and C.9. Since H_x gives objects covered by S_{b_0}, S_{b_1}, $\ldots$, $S_{b_{x-1}}$ and since h (which is formed during the existence iteration) gives the objects covered by $S_{b_{x+1}}$, $S_{b_{x+2}}$, $\ldots$, S_{b_k}, this test asks if there exists an x, $x \in k$, such that S_{b_x} is extraneous by virtue of $H_x \cup h$ covering the same objects as $H_{k+1} = \bigcup_{i=0 \text{ to } k}(S_{b_i})$.

The use of this test to eliminate redundancies is straightforward but rather inefficient since many irrelevant questions are asked (note that redundancy can only occur when $S_{b_x}\cap S_{b_k} \neq \{\ \}$). An apparent improvement (a sacrifice of space for time in medium-sized problems) can be made by precomputing those "redundancy situations" pertinent to each block for use in the impasse detection. Specifically, we precompute subsets of blocks $R_{b_k,i}$, $i = 0$ to $r_{b_k}-1$, which, if contained in C, obviate the inclusion of S_{b_k}. These subsets are derived from covers of a block by other blocks. For instance, if S_{b_0}, S_{b_1}, $\ldots$, $S_{b_{t-1}}$ cover S_p, $p \neq b_h$ for $h \in t$, then for each j, $\{b_0,b_1,\ldots,b_{j-1},b_{j+1},\ldots,b_{t-1},b_p\}$ is an entry of $R_{b_j,\_}$ since if all the indicated blocks are in C, the inclusion of S_{b_j} would create a redundancy. Note that $\{b_0,b_1,\ldots,b_{t-1}\}$ need not be in the $R_{b_p,\_}$ list since, when these indicated blocks are in C, no object of b_p is uncovered and no attempt to adjoin b_p to C will be made.

To compute the redundancy subsets $R_{x,y}$ we must solve m covering problems, namely for each $p \in m$, cover S_p with S_i, $i \in m \sim \{p\}$. The reader will notice that in each of these preliminary problems we again have need for the generation of irredundant covers, hence the complete process is conceptually recursive. In most real problems, however, the number of objects contained in each block is quite small compared to n, and a single recursive step (as outlined above) is sufficient. Such an algorithm is exhibited in Fig. 6.20. The same nest is used to compute preliminary covers ($X = S_p$ for $p \in m$) as final covers ($X = N$ and $p = m$). During the preliminary calculation only basic covers are generated, while during the main calculation (i.e. when $p = m$), the test at line I.12 prevents redundancies from infiltrating and all covers generated are irredundant. The notation is identical to that used in Fig. 6.16 with the addition of the 2-array of redundancy subsets R, the associated 1-array of subset counts r, and the self-explanatory dummies p, X, and i.

6.4.4. *Other Program Refinements*

We discuss here two possible cover generation modifications but leave their precise implementation as exercises (see § 7.5, however).

First, consider the following two possible situations that can occur in an arbitrary

I.1		for $p = 0$ to $m-1$: $r_p = 0$
I.2		for $p = 0$ to $m-1$:
I.3		$X = S_p \cap N$; $G_0 = m \sim \{p\}$; go to #20
I.4	#10	reiterate
I.5		$X = N$; $G_0 = $ (set) m
I.6	#20	$H_0 = \{\ \}$
I.7		with $k = 0, 1, \ldots$, as $C = \{\ \}, \{\ \} \cup \{b_{k-1}\}, \ldots$, until $H_k \supset X$:
I.8		$\cdot$ $\cdot$
I.9		$j = (X \sim H_k)_1$; $B_k = V_j \cap G_k$
I.10		for $b_k \in B_k$:
I.11		$H_{k+1} = H_k \cup S_{b_k}$
I.12		if $p = m$ and $\exists_{i \in r_{b_k}} (R_{b_k, i} \subset C)$: reiterate
I.13		$G_k \sim \{b_k\} \to G_k$
I.14		$G_{k+1} = G_k$
I.15		$\cdot$ $\cdot$
I.16		if $p = m$: USE IRREDUNDANT COVER C
I.17		for $i \in C$:
I.18		$R_{i, r_i} = (C \sim \{i\}) \cup \{p\}$; $r_i + 1 \to r_i$
I.19		if $p < m$: go to #10

FIG. 6.20. An irredundant cover generation program.

incidence matrix: (1) there exists a column with a single nonzero entry, and (2) there exist two rows, one covering the other (i.e. there exist two blocks, one a subset of the other). In situation (1) the lone block containing the singly covered object must be in every cover, hence we may mark this block as "indispensable" and simplify the problem by deleting it and all objects which it covers. In many particular problems (e.g. looking for "minimal" covers), situation (2) also gives rise to a simplification, namely the deletion of the block which is covered (perhaps only if it is a *proper* subset of another block). Now the simplification of situation (1) can give rise to situation (2), as shown in Fig. 6.21(a) and vice versa as in Fig. 6.21(b). The tests and simplifications can be repeated until neither situation occurs. The indispensable blocks so discovered are called a *core* for the problem. (A core is not necessarily unique since there is a choice of simplifications when two identical rows appear.) Often, a core covers all objects and is therefore an irredundant cover. If, however, a core is not a cover, then a branching process as exhibited in Fig. 6.16 can be applied to this reduced problem to produce subsets of blocks which when adjoined to the core yield covers. Of course, a core calculation can be performed at each stage of the branching process since that process itself corresponds to considering mutually exclusive reduced problems (§ 7.5).

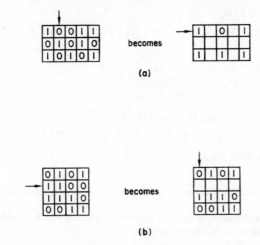

FIG. 6.21. Illustration of incidence matrix reductions associated with "minimal" cover generation.

The second refinement is concerned with the connectivity of a given set of blocks. Suppose that blocks $S_{q_0}, S_{q_1}, \ldots, S_{q_{w-1}}$ have no objects in common with blocks $S_{q_w}, S_{q_{w+1}}, \ldots, S_{q_{m-1}}$, i.e. that $P = \bigcup_{i=0 \text{ to } w-1}(S_{q_i})$ and $Q = \bigcup_{i=w \text{ to } m-1}(S_{q_i})$ are disjoint sets. In this case, covers of P and of Q may be generated independently and combined by the product rule, a cover of $P \cup Q$ being a cover of P adjoined with a cover of Q. In general, there are as many independent problems as there are connected components of the given set of blocks. These components may be computed by the techniques of § 6.3, say by the application of Warshall's method to the incidence matrix $M_{0 \text{ to } m-1}$, where $i \in M_h$ and $h \in M_i$ if and only if $S_i \cap S_h \neq \{ \ \}$. As with core calculation, component detection may be used at each step of the basic branching process since a set of blocks can become disconnected at any stage of the process. Unfortunately, however, the connectivity determination is rather too time-consuming to be used in general in this way.

Further refinement is, of course, possible. An interesting, little-studied issue is the effect upon algorithmic efficiency of the order in which the objects and blocks are considered. Should lightly covered objects be considered before heavily covered objects? Should large blocks covering a particular block be used for branching before small blocks? Our general rules of backtracking (§ 4.2) seem to indicate an affirmative answer to these questions, although the quantitative value of the required program elaborations must, as always, be weighed against the increased complexity of the algorithms.

EXERCISES

1. The simple program of Fig. 6.22 generates covers less efficiently than the program of Fig. 6.16. However, besides its simplicity, it possesses an important property. What is that property?

*2. A careful study of Fig. 6.17 reveals that the impasse detection (6.5) can be meaningfully applied to *general* cover generation.

L.1 $b_0 = -1; H_0 = \{\ \}$

L.2   with $k = 0, 1, \ldots$, as $C = \{\ \}, \{\ \} \cup \{b_k\}, \ldots$, until $H_k \supset N$:

L.3 $\ddots$

L.4   for $b_{k+1} = b_k + 1, b_k + 2, \ldots, m - 1$:

L.5 $H_{k+1} = H_k \cup S_{b_{k+1}}$

L.6 $\ddots$

L.7   for $D \subset m \sim (b+1)$:

                USE $C^* = D \cup C$ AS A COVER

FIG. 6.22. An alternative basic cover generation program.

(a) Study, empirically, the effect of the incorporation of this impasse detection on the basic generation of Fig. 6.16.

(b) Incorporate impasse detection of this sort into the program of Fig. 6.22 and compare the efficiency of the resulting program with that of part (a).

3. In what way may the following incidence system be simplified prior to cover generation? Draw the search tree corresponding to the generation of basic covers via the program of Fig. 6.16 for the simplified system.

l		l	l		l		
l				l	l	l	
		l		l			l
		l			l		
l			l	l	l	l	

4. Find a Steiner triple system for $n = 13$.

5. Modify the program of Fig. 6.19 so that it will compute, for *arbitrary n*, the cardinality of the largest set of triples of labels $0, 1, \ldots, n-1$ such that no pair of labels appears in more than one triple. (For an analytic solution to this problem see Spencer [186].)

6. Write a program to calculate a core for an arbitrary incidence matrix using the simplifications discussed in § 6.4.4 iteratively.

*7. (a) Write a program to calculate the components of blocks for an arbitrary incidence matrix

   (b) Modify the program of Fig. 6.16 so that it generates covers for a given component of blocks $K, K \subset m$.

   (c) Incorporate the programs of parts (a) and (b) as procedures in a general cover generation procedure.

*8. Calculate the number of 5×5 zero–one matrices whose permanent is zero. (Hint: If a row contains only zeros, the permanent is certainly zero since every term of the permanent is *covered*.)

9. For a given incidence system consisting of objects $\{0, 1, \ldots, n-1\}$ and blocks B_0 to $m-1$, a system of distinct representatives (an SDR) is a vector $(r_0, r_1, \ldots, r_{m-1})$ with distinct components such that $r_i \in B_i$ for $i \in m$. Write an efficient program to generate all SDR's for a given incidence system. (Note: A theorem of P. Hall (see M. Hall [166], for instance) says that an SDR exists if and only if

$$\forall_{K \subset m} (|\bigcup_{k \in K} (B_k)| \geqslant |K|)$$

is true. An SDR is the same as a *complete matching* in a bipartite graph—see Liu [28].)

10. Write a program to generate the ways in which k nonattacking rooks can be placed on the triangular chessboard

The number of ways is the Stirling number of the second kind $S_{q+1,q+1-k}$ —see Appendix I.

6.5. Transformations

Transformations of finite sets, hence of the objects which they model, play a significant role in combinatorial computing. In this section we discuss certain common elementary transformations. Also, we present a scheme for generating a (small) *group* of permutations from an arbitrary *set* of permutations. The important subject of the equivalence of configurations under a given group of transformations is pursued in § 6.6.

6.5.1. *Matrix Translation and Transposition*

The simplest Euclidean motion is translation, moving an object in a particular direction without rotation, expansion, or contraction. In one dimension, translation "to the right" (negative motion) or "to the left" (positive motion) is indicated for sets of whole numbers by the $\ominus$ and $\oplus$ operations respectively. Translation in n dimensions is specified by describing the positive or negative motion in each of n orthogonal directions. Notationally, this motion is indicated by n subscript transformations as in

$$\text{for } i = 4 \text{ to } 11; \quad j = 10 \text{ to } 0: A_{i,j} \rightarrow A_{i-4,j+5} \tag{6.6}$$

for example. This translation is presented pictorially in Fig. 6.23. Note the order specified for the indices in these examples. Since the translated array may overlap the original array, it is necessary to move first those entries which may later be covered up. This is

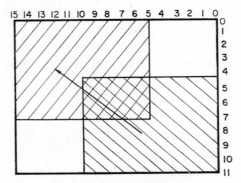

FIG. 6.23. Matrix translation.

accomplished by iterating backwards in dimensions which have positive translations and forwards in dimensions which have negative translations.

Another common transformation is matrix *transposition*, interchanging the rows with the columns of a given 2-array. If $A_{1 \text{ to } I, 1 \text{ to } J}$ is the given matrix, then $B_{1 \text{ to } J, 1 \text{ to } I}$, where $B_{j,i} = A_{i,j}$ for $i = 1$ to I and $j = 1$ to J, is the *transpose* of A. If separate storage is used for B, then this transposition may be effected directly, as indicated above, or by the following program when A is a 2-array of zeros and ones viewed as a 1-array of sets:

$$\text{for } j = 1 \text{ to } J\colon B_j = \bigcup_{i=1 \text{ to } I \ni j \in A_i} (\{i\}). \tag{6.7}$$

(Transposition of a zero–one matrix is an operation deserving investigation by hardware designers. In many problems it is necessary to preserve both a zero–one matrix and its transpose in order to obviate the use of a time-consuming transpose program.)

Often it is inconvenient or inefficient to use completely disjoint storage for the transpose. A common manipulation is to transpose an I by J 2-array A by transposing a $K \times K$ square matrix M, $K = max(I,J)$, which "contains" A. In this case the entries symmetric about the main diagonal of the square matrix M are interchanged. For instance, the iteration

$$\text{for } i = 1 \text{ to } K; \quad j = i+1 \text{ to } K\colon M_{i,j} \leftrightarrow M_{j,i} \tag{6.8}$$

transposes the square matrix $M_{1 \text{ to } K, 1 \text{ to } K}$. If M is a 1-array of sets given in bit-pattern representation, then the body of this loop could be replaced by

$$
\begin{aligned}
&b_i = M_i \cap \{j\}; \quad b_j = M_j \cap \{i\} \\
&(M_i \sim \{j\}) \cup (b_j \oplus \{j-i\}) \to M_i \\
&(M_j \sim \{i\}) \cup (b_j \ominus \{j-i\}) \to M_j
\end{aligned}
\tag{6.9}
$$

for example.

The problem of transposing an $I \times J$ matrix onto itself (i.e. using the identical storage) is an interesting combinatorial problem considered by various authors: Berman [154], Windley [194], and Pall and Seiden [178]. One wishes to transpose a matrix $Q_{0 \text{ to } I-1, 0 \text{ to } J-1}$, which is stored as $C_{0 \text{ to } IJ-1}$, merely by permuting entries—the transpose ultimately occupying the same storage as did Q. Since $C_{iJ+j} \equiv Q_{i,j}$, the problem consists of effecting the permutation defined by the transformation $C_{jI+i} \to C_{iJ+j}$. Since, for $i \in I$ and $j \in J$, we have that $J \times (jI+i)$ modulo $IJ-1$ equals $iJ+j$ (except when $i = I-1$ and $j = J-1$), the permutation of the subscripts of C may be defined by

$$
\begin{aligned}
&p_0 = 0; \ p_{IJ-1} = IJ-1 \\
&\text{for } k = 1 \text{ to } IJ-2\colon \\
&\quad p_k = kJ \ (\text{mod } IJ-1)
\end{aligned}
\tag{6.10}
$$

Thus, to accomplish the transposition, we may apply the program of Fig. 6.8 with $N = IJ$ and $p_{0 \text{ to } N-1}$ precomputed from (6.10).

6.5.2. *Symmetries of the Square—The Dihedral Groups*

Transposition of a square matrix is equivalent in effect to reflection of the matrix about its main diagonal. This is but one of the eight distinct rigid motions which map a square onto itself. These transformations are often applied to square matrices stored within a computer in order to detect symmetries of two-dimensional figures composed of nonzero entries of the matrix; for example, these transformations are required in manipulations with polyominoes.

Let I represent the identity transformation, D represent reflection about the main diagonal (transposition), V represent reflection about a vertical axis, H represent reflection about a horizontal axis, and R represent 90° clockwise rotation. The group of the square may be generated by either $\{D,R\}$ or $\{D,V,H\}$ as illustrated in Fig. 6.24. The

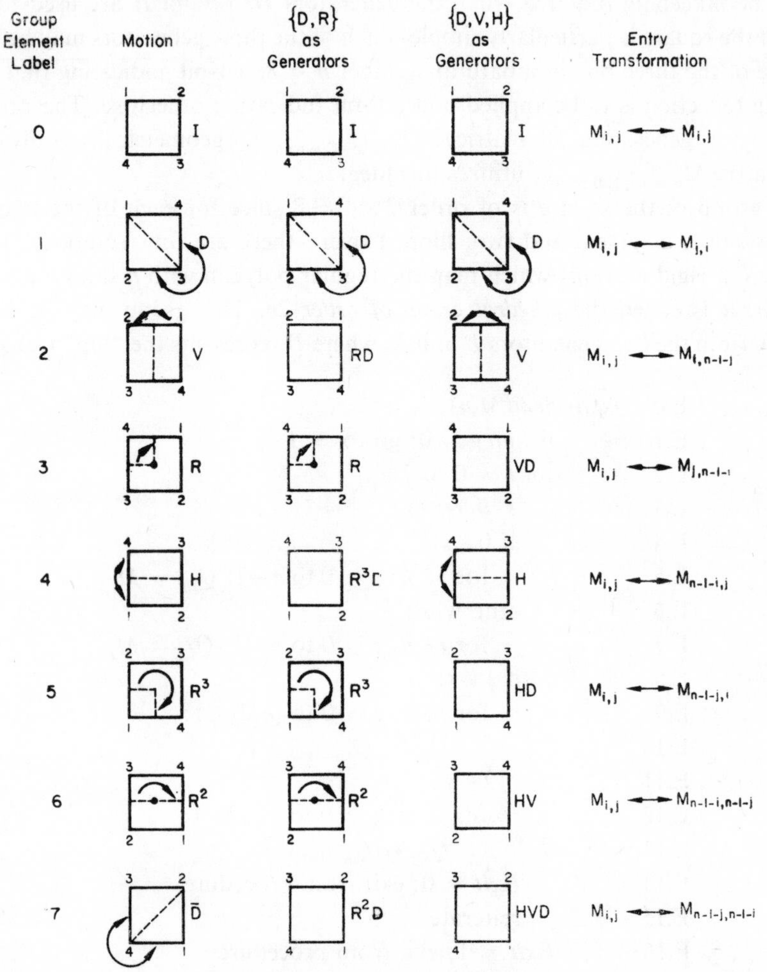

FIG. 6.24. The group of the square.

movement induced upon a typical entry $M_{i,j}$ of a matrix $M_{0 \text{ to } n-1, 0 \text{ to } n-1}$ by the corresponding group element is given in the rightmost column of the figure.

Note that all transformations of a matrix onto itself except those labeled 3 and 5 may be accomplished with entry interchanges (about $n^2/2$ of them), hence the structure of programs (6.8) and (6.9) apply. The 90° and 270° rotations are more involved since the cycles of the permutation are of length four instead of two. While this may be done in a manner similar to that illustrated in Fig. 6.8 (the use of the mask S is unnecessary in this special case, however), it is conceptually simpler to use the generators D, V, and H—i.e., D followed by V, $V \cdot D$, and D followed by H, $H \cdot D$—to effect these two transformations. Of course, when separate storage is used for the transformed matrix, "overwriting" problems do not arise and each of the eight motions may be programmed directly—see (6.7), for instance.

The bookkeeping required when the generators D, V, and H are used to form the group of the square is particularly simple. Each of the three generators may be associated with one of the three bits of a natural number $n < 8$, a 1-bit indicating that the corresponding reflection is to be applied and a 0-bit indicating otherwise. The procedure of Fig. 6.25 for generating all matrices $Q_{0 \text{ to } n-1, 0 \text{ to } n-1}$ "geometrically equivalent" to a given matrix $M_{0 \text{ to } n-1, 0 \text{ to } n-1}$ utilizes this idea.

The group of the square is of order $2 \times 4 = 8$ since for each of the two reflective positions—one, as given, and two, flipped over—there are four rotations. In general, the group of rigid motions which map the regular polygon with n sides onto itself is of order $2n$; it is called the *dihedral group of order* $2n$. This group may be constructed naturally from the two generators D and R, where D represents the "flip" transformation

```
E.0     equivalent(M,n)
E.1          if Entry = 0: go to # 1
E.2          for t = 0 to 7:
E.3              s = set(t)
E.4              if 0 ∈ s:
E.5                  for i ∈ n; j = 0 to n−1: Q_{j,i} = M_{i,j}
E.6              otherwise:
E.7                  for i ∈ n; j = 0 to n−1: Q_{i,j} = M_{i,j}
E.8              if 1 ∈ s:
E.9                  for i ∈ n; j = 0 to [n/2]−1:
E.10                     Q_{i,j} ↔ Q_{i,n−1−j}
E.11             if 2 ∈ s:
E.12                 for j ∈ n; i = 0 to [n/2]−1:
E.13                     Q_{i,j} ↔ Q_{n−1−i,j}
E.14             Exit = 0; exit from procedure
E.15    # 1      reiterate
E.16         Exit = 1; exit from procedure
```

FIG. 6.25. A program to generate the group of the square.

(reflection) and R represents the rotation through $360/n$ degrees. For example, if the vertices of the polygon are labeled 0, 1, $\ldots$, $n-1$, then the dihedral group is generated by the two permutations $(n-1,n-2,\ldots,0)$ and $(1,2,\ldots,n-1,0)$. The dihedral groups arise in investigations of planar graphs—see § 7.3.

6.5.3. *Permutation Group Generation*

In many combinatorial problems which involve a group of transformations, the actions of the transformations are built into the program. Occasionally, however, it is convenient to precompute the transformations of the group, storing them explicitly as permutations. For common groups such as the symmetric group, alternating group, dihedral group, or group of the N-cube (see exercise 5), this computation presents no particular problem (e.g. see § 5.2). We now consider the more general problem of generating the smallest group which contains a given arbitrary set of transformations.

G.1	for each *untested*$(G^* \sim G) \rightarrow g^*$:
G.2	$a = numtoperm(g^*,n)$
G.3	$G \cup \{g^*\} \rightarrow G$
G.4	for $g \in G$:
G.5	$b = numtoperm(g,n)$
G.6	for $i = 0$ to $n-1$: $p_i = a_{b_i}$
G.7	$G^* \cup \{permtonum(p)\} \rightarrow G^*$
G.8	for $i = 0$ to $n-1$: $p_i = b_{a_i}$
G.9	$G^* \cup \{permtonum(p)\} \rightarrow G^*$
G.10	"G (or G^*) is the new group."

U.0	*untested*(K)
U.1	$k = (K)_1$
U.2	if nonexistent: $Exit = 1$
U.3	otherwise: $Exit = 0$; $untested(\,) = k$
U.4	exit from procedure

FIG. 6.26. Formation of the group encompassing a given set of permutations.

A program to solve this problem is shown in Fig. 6.26. We are initially given a set of transformations G^* and a subset G of G^* known to constitute a group (possibly containing only the identity element). In this program the transformations are represented by serial numbers of permutations of $(0,1,\ldots,n-1)$ so that sets of permutations are represented by subsets of $\{0,1,\ldots,n!-1\}$. (When n is large, as it often is, this representation must be replaced by the list representation—see exercise 7.) The idea of the algorithm is to enlarge G within G^* until it is closed under multiplication. After admission to G (at line G.3), an element of $G^* \sim G$ is tested in products with every element of G (via the loop at lines G.4 through G.9). These products, some new and some old, are adjoined to G^* (at lines G.7

and G.9). Since the eventual group is finite, sooner or later no new elements will appear (and G becomes equal to G^*).

This algorithm is slow, particularly when G is initially small compared to the final group, since so much time is spent in forming products which already exist in G^*. For instance, if G were initially equal to $\{0\}$ and G^* were a group of order m, it would still take nearly $m(m-1)/2$ passages through the time-consuming program at lines G.5 through G.9 to establish G^* as a group. Thus this algorithm is practical only when the number of new group elements is reasonably small (say less than 100). Significant improvements of this situation depend upon group-theoretic analysis which we do not pursue in this elementary exposition (see exercise 8).

EXERCISES

1. Describe the motion indicated by the transformation $A_{i,j} \rightarrow A_{3i+1,j/2-1}$. Write a loop index generation procedure which will calculate a correct sequence of values for i and j to accomplish this transformation when the range of i is 0 to I and the range of j is 2, 4, ..., $2J$.

2. Write a program to calculate the inverse of a permutation $p_{0 \text{ to } n-1}$ "in place" (i.e. without using an additional array to hold the inverse). (Hint: The transpose of a permutation matrix is the permutation matrix for the inverse permutation.)

3. Write an efficient program for performing a 90° clockwise rotation of a square matrix in place.

4. Calculate the number of solutions to the queens' problem for $n = 4$ through 13 which are inequivalent under the group of the square.

5. There are $2^N N!$ transformations of the N-cube into itself.
   (a) List the movements induced upon the cubical array entry $M_{i,j,k}$ by the forty-eight symmetries of the 3-cube.
   (b) With the 2^N vertices of the N-cube labeled with the integers 0, 1, ..., $2^N - 1$ in the natural fashion (see § 3.5), write a program to generate the $2^N N!$ vertex permutations which describe the symmetries of the N-cube.

6. Write a program which constructs the permutations of the cyclic group generated by an arbitrary permutation of $(0,1,\ldots,n-1)$.

7. Modify the program of Fig. 6.26 as dictated by use of lists of permutations as vectors instead of by use of sets of serial numbers. In this respect see the Algol algorithm [192] by M. Wells (not this author).

*8. It can be shown [12] that the two permutations $(2,3,\ldots,n,1)$ and $(2,1,3,4,\ldots,n)$ generate the symmetric group of order n. Use this criterion to improve the group generation algorithm presented in § 6.5.3.

6.6. Isomorph Rejection

The concepts of isomorphism and symmetry, so important in all of mathematics, pervade combinatorial computing to an exceptional degree. Not only are counting equivalence classes, selecting representative configurations, etc., inherent in many combinatorial problems, but the elimination of repetitious computation by the detection of isomorphic structure is often essential to the construction of practical algorithms.

In this section we discuss the important [189] problem of the generation of configurations inequivalent under a prescribed group of transformations. We assume that a generation scheme for *unrestricted* configurations is known. The group, which usually arises from a consideration of natural transformations between the unrestricted configurations, defines an equivalence relation on these configurations. The problem then is to generate, or perhaps just select, a representative from each resulting equivalence class.

6.6.1. *A Typical Problem—Basic Program Structures*

As a typical problem of isomorphic rejection, consider the job of finding representative k-subsets of the n-cube under its basic symmetry group. That is, we wish to find k-subsets of $\{0,1,\ldots,2^n-1\}$ which are inequivalent under simultaneous permutation of the n-bits of the elements written in binary (rotation) and/or independent interchange of the 0-bits with the 1-bits in any bit position of the elements (reflection). This problem may be viewed as the generation of representative $k \times n$ zero–one matrices with the rows ordered, where column permutations and column complementations are the allowable equivalence transformations. The group has cardinality $2^n n!$ and is applied to $\binom{2^n}{k}$ unrestricted configurations. For $n = 3$, the fifty-six 3-subsets fall into three equivalence classes which may be represented by the subsets $\{000,001,010\}$, $\{000,011,101\}$, and $\{000,001,110\}$.

To understand the various ways in which these representatives may be generated, consider the arborescent generation of all k-subsets as effected by segment N of Fig. 5.6. This generation is diagrammed, for $n = 3$ and $k = 4$, in Fig. 6.27. If the complete tree were drawn there would be 1680 terminal nodes; however, we have not drawn branches.

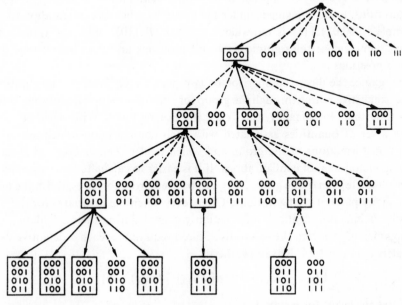

FIG. 6.27. The search tree for the generation of representative 4-subsets of the 3-cube.

which emanate from nodes associated with partial subsets equivalent under the group to previously generated partial subsets.

For such an arborescent generation, the rejection of redundant isomorphic computation is structurally identical to impasse detection. In fact, the possible structures of isomorph rejection programs arise from the various ways in which this impasse detection is inserted into the generation. Generally speaking, there are three basic structures for these programs. These are based on three different ways in which partial configurations may be tested for equivalence: by sieving, by direct comparison, and by indirect comparison.

In § 6.1.2 we discussed the sieving approach as applied at the bottom of the tree. In general, one may sieve at other levels of the generation rather than solely after the entire generation is complete. Sieving is relatively simple but only practical when neither the cardinality of the group nor the number of (partial) configurations is excessively large. It is useful when in addition to the representatives one wishes to know the number of configurations in each equivalence class.

6.6.2. *Direct Comparison*

In the direct comparison approach, one saves representative partial configurations at each level for comparison with a newly generated partial configuration. It is best for sake of economy of storage to use the tier scan program structure discussed in § 4.4—see especially Fig. 4.23. In the illustrated generation for our typical example (Fig. 6.27), the representatives at each level are boxed. Actually, all configurations shown would be generated, but those at the end of a dotted arrow would be found, by a "direct" comparison of configurations, to be equivalent under the group to a boxed configuration to its left. For example, {000,011,111} is equivalent to {000,001,110} under the transformation which says complement the rightmost two bit positions and then interchange the first and last bit positions.

At first glance the direct comparison of two possibly equivalent configurations involves the exhaustive search through the group for a transformation taking one configuration into the other. Fortunately, this is usually not required. With ingenuity one can often find a set of quantities associated with each configuration which are invariant under all transformations of the group and which tend to characterize small sets (one hopes for singleton sets) of configurations. If it is easier to calculate these *invariants* than to make a complete scan of the group, they can be used profitably to establish the inequivalence of many pairs of configurations. Often, even when the invariants for the comparand configurations are identical, information gathered during their calculation can be used to significantly shorten the exhaustive search required to reach a definitive decision on equivalence. In our typical example the set

$$\bigcup_{i \in h; j \in h \sim (i+1)} \left(\{ |\, set[(s)_{i+1}] \cap set[(s)_{j+1}] | \} \right) \tag{6.11}$$

is an effective invariant for h-subsets s.

The direct comparison approach is applicable in a large number of problems and must be considered the primary means of isomorph rejection. Since its chief drawback is the storage required for the many complex configurations, it is especially appropriate when the group is large and therefore the number of representatives is small. Further discussion including a specific example of this method appears in § 7.1. Additional discussion of invariants may be found in Golomb [17].

6.6.3. *The Indirect Comparison Approach*

The third basic means of isomorph rejection is the indirect approach. In its full generality it is practical only in special cases (see exercise 4, for instance), but various watered-down versions are often useful for impasse detection in covering problems. Roughly speaking, whereas the direct approach prevents repeated computation by saving representatives, the indirect approach sacrifices time for storage.

In this approach the arborescent generation of unrestricted configurations must always proceed straightforwardly so that the nest vector (or other easily calculated ancestral specification) changes value in a strictly monotone fashion. At each node a test is made to see if there exists a transformation carrying the partial configuration into one with a, say, smaller nest vector. If the test is successful, then this partial configuration has been formed, in disguise, earlier in the generation, and backtracking is initiated. If the (exhaustive) test fails to find a transform with a smaller nest vector, then this partial configuration is the first representative of an equivalence class at the particular tree level and the generation proceeds deeper. The diagram of Fig. 6.27 illustrates such a generation: backtracking occurs at the end of each dotted arrow while deeper nesting occurs at the end of a solid arrow (until the bottom is reached, or an impasse of another sort arises).

This test for equivalence can be implemented in various ways. One way is simply to calculate the nest vector of each configuration transform to compare with the configuration nest vector. However, since many group elements give rise to the same calculation order for the resulting configuration, hence to the same nest vector (especially during early stages of the basic generation), it is usually best to invert this calculation. That is, one examines other ways in which the partial configuration could be formed, allowing for transformations under the group, continually comparing the nest vector of these alternatives with the nest vector of the partial configuration.

We give in Fig. 6.28 a complete program for our typical problem using the indirect approach of equivalence testing. Segment B is the basic generation as in segment N of Fig. 5.6. The nest vector $(f_1, f_2, \ldots, f_k)$ itself provides the generated configurations. Thus the equivalence test (line B.5 and segment E) consists of a search for a permutation $(p_1, p_2, \ldots, p_h)$ and a transformation t of the group such that

$$(t(f_{p_1}), t(f_{p_2}), \ldots, t(f_{p_h})) < (f_1, f_2, \ldots, f_h).$$

The permutation generation structure (the nest at lines E.3 through E.17) is familiar, only here one level at the outside of the nest has been distinguished (the loop whose

B.1 $f_0 = -1$

B.2   with $h = 1$ to k:

B.3 $\ddots$

B.4   for $f_h = f_{h-1}+1, f_{h-1}+2, \ldots, 2^n-1$:

B.5   if $equivalence(\ ) = 1$: reiterate

B.6 $\ddots$

B.7   USE $\{f_1, f_2, \ldots, f_k\}$

B.8   stop

E.0 $equivalence(\ )$

E.1   for $p_1 = 1$ to h:

E.2   if $0 < f_1$: $equivalence(\ ) = 1$; exit from procedure

E.3 $S = (h+1) \sim \{0, p_1\}$; $G_1 = $ (set) $n!$

E.4   with $i = 2$ to h:

E.5 $\ddots$

E.6   for $p_i \in S$:

E.7 $b = set(f_{p_i}) \triangle set(f_{p_i})$; $G_i = \{\ \}$

E.8   for $g \in G_{i-1}$:

E.9 $q = numtoperm(g, n)$

E.10 $b^* = \bigcup_{j \in b} (\{q_j\})$

E.11   if $nbr(b^*) < f_i$:

E.12 $equivalence(\ ) = 1$; exit from procedure

E.13   if $nbr(b^*) = f_i$: $G_i \cup \{g\} \to G_i$

E.14   if $G_i = \{\ \}$: exit from loop

E.15 $S \sim \{p_i\} \to S$

E.16 $\ddots$

E.17 $S \cup \{p_i\} \to S$

E.18 $equivalence(\ ) = 0$; exit from procedure

Fig. 6.28. A program for the indirect approach to isomorph rejection.

quantifier is at line E.1). This simplifies handling of the group since choice of the first subset element establishes the "parity" of each bit position, hence only permutation of the bit positions need be considered within the nest. As successive subset elements are chosen (line E.6), the subgroup which leaves the partial subset invariant is calculated by the iteration at lines E.8 through E.13. During this calculation, if a group element is found which yields a smaller partial subset (i.e. if the test at line E.11 is successful), then equivalence with a previously generated subset is established. If the subgroup is void (line E.14), then backtracking within the equivalence test is possible since all further generated permutations have nest vector (i.e. partial subset) exceeding $(f_1, f_2, \ldots, f_h)$.

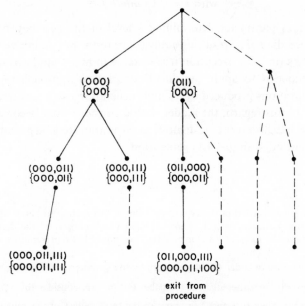

FIG. 6.29. Illustration of the equivalence search for the indirect approach.

A diagram of the execution of this equivalence test for the partial subset (000,011,111) appears in Fig. 6.29. The entire permutation is sketched although only the solid links would actually be followed. The vectors given at each node represent the permutation of (000,011,111) being generated. The sets written below these vectors are the smallest existing transforms of these vectors. The link emanating from the {000,111} node is not followed since (000,111,—) > (000,110,111) and the test at line E.14 would be successful. Since (000,011,100) < (000,011,111), the equivalence test itself is successful when that node is reached.

Many variations of this illustrative program are possible. For instance, one can split off the second nest level in addition to the first. By precomputing the "invariant subgroups" associated with each possible second permutation component, the group iteration at lines E.8 through E.13 can be eliminated for $i = 2$, the time at which it is especially long-winded. It should be pointed out that an invariant subgroup *precomputation* effectively permits the basic configuration generation itself to be shortened. For example, in Fig. 6.28 there is really no need to generate subsets with smallest element greater than zero (i.e. $f_1 > 0$) since these subsets are known to be equivalent to a previously generated subset. (If the generation were so modified, the test at line E.2 could of course be omitted.) This variation is pursued further in exercise 6. Furthermore, in many cases, the group iteration can profitably be replaced by a nest generation. Such a generation is usually more efficient since it may contain the "large vector" test (line E.14) as impasse detection *within* itself.

Another variation is to replace the nest quantifier at line E.4 with the quantifier

$$\text{with } i = 2 \text{ to } min(h,L): \tag{6.12}$$

where L is an input parameter determining a level of the nest beyond which there is little point in proceeding. The resulting equivalence test is not definitive, but since it does consume much less time for execution it is suitable in many applications. For instance, a common technique is to apply the direct comparison approach to the manageable number of configurations produced by a simple generation using a watered-down indirect equivalence test. In this regard, the reader should consider what becomes of the indirect approach when the arborescent configuration generation (e.g. segment B of Fig. 6.28) is replaced by a cut-off scan (§ 4.4.1) generation.

EXERCISES

1. Consider the generalization of the typical problem of this section to bases other than 2. That is, consider representative k-subsets of $\{0,1,\ldots,b^n-1\}$ under permutation of the digit positions and/or independent permutation of $(0,1,\ldots,b-1)$ in any digit position. Find representative 4-subsets for $b = 3$ and $n = 2$.

*2. Consider the group of order $48\cdot64$ generated by the symmetries of the 3-cube and independent interchange of diagonally opposite edges of the cube. For $m \in 13$, consider the $3^m \binom{12}{m}$ configurations defined by assigning the labels 0, 1 and 2 to m out of the twelve edges of the cube in all possible ways (the remaining edges remain unlabeled). Write a program to calculate, for $m = 0$ to 12, representative configurations inequivalent under the group *and* the number of configurations represented.

3. Find a pair of inequivalent partial subsets (for the typical example) for which (6.11) fails to discriminate.

*4. Calculate, for $N \le 5$, the number of Hamiltonian cycles of the N-cube inequivalent under the $2^N N!$ symmetries of the N-cube. Gilbert [162] showed that there are nine such cycles for $N = 4$; for $N = 5$ there are more than 200,000. (Hint: The indirect approach is feasible here since there are so few ways, namely 2^{N+1}, in which a given cycle may be formed.)

*5. Suppose that n two-team games (e.g. football games) are to be played with each game having one of three possible outcomes: team A wins, team B wins, or tie. An interesting question, studied by various authors (see Kamps and van Lint [171]), is the minimum number of forecasts—predictions on the outcome of all n games—necessary to insure that, for at least one forecast, at most one game is predicted incorrectly. Write a program which could be used to show that for $n = 5$ this number is exactly 27. (Note: This problem is basically a covering problem; isomorph rejection is required to reduce the very large number of possibilities to manageable proportions.)

6. There are k possible two elements "starts" for generating representative k-subsets of the n-cube under its symmetry group. That is, the first element may always be zero and the second element is determined by the number of 1-bits in its binary expansion, of which there are k possibilities.

(a) Modify (both segments of) the program of Fig. 6.28 to accommodate this preanalysis.

(b) Compare, empirically, the efficiency of this modified program with the basic program of Fig. 6.28.

7. Modify the program of Fig. 6.28 with the iteration over the group (lines E.8 through E.14) replaced by an efficient nest generation of the group.

*8. Calculate the number of Steiner number triple systems for $n = 19$ which are inequivalent under arbitrary label permutations. For $n = 15$, there are eighty such systems—see White, Cole, and Cummings [193] and Hall and Swift [167]. (As far as the author is aware, the number for $n = 19$ is not yet known.)

*9. Write a program to calculate the number of n-posets. These numbers for $n = 4$, 5, and 6 are given on page 4 of Birkhoff [155].

CHAPTER 7

APPLICATIONS—ADVANCED ALGORITHMS

CHAPTERS 3 through 6 are chiefly concerned with abstract combinatorial manipulations. In this chapter we delve more deeply into the manner in which these manipulations are used in real problems. A few important algorithms are indeed discussed in detail (§§ 7.1 and 7.2.3, for instance), but the emphasis here is on the framework of involved combinatorial investigations and their synthesis from elemental algorithms.

The problems described in this chapter are ones with which the author has had intimate contact. They represent but a small fraction of the innumerable combinatorial studies that are amenable to the techniques presented in this book. The bibliography for this chapter contains references to many such investigations.

7.1. Incidence Matrix Equivalence

In most incidence systems which arise in combinatorial computing, the labeling of the blocks and of the objects is entirely arbitrary. Thus it is often necessary, or desirable, to detect the equivalence of two zero–one matrices under independent permutation of the rows and columns. This important general problem is now discussed in conjunction with a characteristic application—the enumeration of reduced Latin squares.

7.1.1. *Invariants*

Let X and Y be two incidence matrices which we wish to test for equivalence. The basic approach used here is to determine whether there exists an association of X-rows to Y-rows and X-columns to Y-columns such that if the corresponding row and column permutations were actually effected, X would become identical to Y. While the generation and trial of *all* permutations is usually unthinkable, it is often possible to prune the generation sufficiently to make this approach feasible.

This pruning is based upon the calculation of certain invariant weights for the rows and columns of X and Y. If X is indeed equivalent to Y, each row or column of X has the same weight as its associated row or column in Y. Thus the permutation generation may be restricted to the association of equally weighted rows and columns. Indeed, the set of row weights and the set of column weights are invariants which may establish the inequivalence of the given matrices.

To discuss a set of weights and invariants which the author has found to be both easily calculable and effective in varied applications, it is convenient to treat the rows and columns on an equal footing. We therefore assume that both the rows and columns of X and Y are given as 1-array of sets, $Rx_{0 \text{ to } B-1}$, $Cx_{0 \text{ to } J-1}$ for X, and $Ry_{0 \text{ to } B-1}$, $Cy_{0 \text{ to } J-1}$ for Y; B is the number of blocks (rows) and J is the number of objects (columns). For $i \in B$ and $k \in B$, let $Drx_{i,k} = |Rx_i \cap Rx_k|$. These quantities estimate the correlation between pairs of rows. Now for each i, let Prx_i be the partition associated with the composition $(Drx_{i,0}, Drx_{i,1}, \ldots, Drx_{i,B-1})$. This partition (into B parts with zeros allowed, no part exceeding $|Rx_i|$) measures the correlation of row i to all rows of X without regard to any ordering of the rows. The weight of row i, Wrx_i, is simply a (the) serial number of Prx_i. The vector $Irx = (Wrx_{r_0}, Wrx_{r_1}, \ldots, Wrx_{r_{B-1}})$, with $Wrx_{r_0} \leqslant Wrx_{r_1} \leqslant \ldots \leqslant Wrx_{r_{B-1}}$, is an invariant, which we call the *row-invariant*, of the matrix X. We analogously define the column weights Wcx_i, $i \in J$, and the column-invariant Icx.

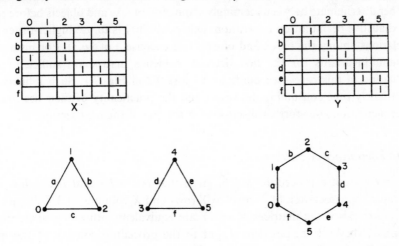

Irx, Iry, Icx, Icy = $(serial(Prx_a), \cdots, serial(Prx_f))$, where Prx_a = (2,1,1,0,0,0), etc.

Irx, Icx = $(serial(Prx_A), \cdots, serial(Prx_F))$, where Prx_A = (3,3,3,0,0,0), etc.

Iry, Icy = $(serial(Pry_A), \cdots, serial(Pry_F))$, where Pry_A = (4,3,3,2,2,2), etc.

FIG. 7.1. Example of certain incidence matrix invariants and their effectiveness.

These weights and invariants are easily calculated using techniques presented else-where in the book. The sorting of the Dr's to produce a partition is efficiently handled by Shell's method discussed in § 6.2.2. The serial number calculation reduces to *comb-tonum*( ) by virtue of exercise 20 of § 5.5; although, in special cases many of these partitions cannot exist and use of an indirect scheme (exercise 4 of § 6.2) is desirable. It is efficient to use a fractional word representation for the invariant ordered B-tuples and J-tuples (Irx and Icx) so that their comparison is fast.

If X is equivalent to Y, then $Icx = Icy$. The truth of these equalities, however, is not sufficient to insure equivalence, i.e. these invariants are not a *complete set of invariants*. A simple example of inequivalent matrices with identical row- and column-invariants is given at the top of Fig. 7.1. Here $Drx_{0,0} = 2$, $Drx_{0,1} = 1$, etc., and $Prx_a = (2,1,1,0,0,0)$. Also, $Prx_b = \ldots = Prx_f = (2,1,1,0,0,0)$; in fact all of the row and column partitions for both X and Y (all the Prx's, Pcx's, Pry's and Pcy's) equal this same partition. Thus all the invariants are identical.

Depending upon the particular application, one may wish to refine these invariants and further discriminate between seemingly symmetric blocks and objects before resorting to the exhaustive permutation generation. One possibility is to use higher-order connec-tivity relations among the block and objects (see exercise 3 of § 6.3). For instance, row- and column-invariants for the first derived incidence matrix are sufficient to effect discrimination in this case—see matrices $X2$ and $Y2$ in Fig. 7.1. Such refinements are often extremely time-consuming, however, and the practicality of their use is a matter for experimentation—see further discussion at the end of the next section.

7.1.2. *An Equivalence Algorithm*

We now present a procedure which, given that $Irx = Iry$ and $Icx = Icy$, searches for a compatible association of X-rows to Y-rows and X-columns to Y-columns. If such an association exists, the matrices X and Y are equivalent; otherwise, since the search is exhaustive, they are inequivalent. Input to the procedure consists of the matrices, given by the 1-arrays Rx, Cx and Ry, Cy, and two 1-arrays $Er_{0 \text{ to } B-1}$ and $Ec_{0 \text{ to } J-1}$ where

$$\text{for } i \in B: Er_i = \bigcup\nolimits_{h \in B \ni Wrx_i = Wry_h}(\{h\})$$

and

$$\text{for } i \in J: Ec_i = \bigcup\nolimits_{h \in J \ni Wcx_i = Wcy_h}(\{h\}).$$

The set Er_i contains indices of those rows in Y which have the same weight as $X_{i, -}$, hence which might be equivalent to row i under a permutation; $Ec_{0 \text{ to } J-1}$ are analogous column index sets. Exercise 5 discusses a useful refinement to these quantities.

The complete process is presented in Figs. 7.2 and 7.3 as the co-procedures *equiva-lence*( ) and *association*( ). The idea is to find associations of X-rows to Y-rows and X-columns to Y-columns which are consistent with the input. The sets Bx and By record the row indices for X and Y respectively for which an association has been made; the sets Jx and Jy are the analogous column indices. An association for one row is made by

E.0   *equivalence*(*J,B,Rx,Ry,Cx,Cy,Er,Ec*)

E.1   (real) *J, B; Bx, By, Jx, Jy* = { }

E.2   for $i \in B \sim Bx$:

E.3   for $R_i \in Er_i \sim By$:

E.4 $Bx \cup \{i\} \to Bx$; $By \cup \{R_i\} \to By$

E.5 $S_0 = \{i\}$

E.6   with $h = 1, 2, \ldots$, until $S_{h-1} = \{ \ \}$:

E.7 $\cdot \ \cdot$

E.8   "*P*" = "*R_*"; "*Q*" = "*C_*"

E.9   for each *association*(S_{h-1};"*Ec*","*Jx*",

E.10   "*Jy*","*Rx*","*Ry*") $\to V_h$:

E.11   "*P*" = "*C_*"; "*Q*" = "*R_*"

E.12   for each *association*(V_h; "*Er*",

E.13   "*Bx*","*By*","*Cx*","*Cy*") $\to S_l$:

E.14 $\cdot \ \cdot$

E.15   go to $\#$ 10

E.16 $Bx \sim \{i\} \to Bx$; $By \sim \{R_i\} \to By$

E.17   *equivalence*( ) = 0; exit from procedure

E.18 $\#$ 10   reiterate

E.19   *equivalence*( ) = 1; exit from procedure

FIG. 7.2. Part I of a *direct* incidence matrix equivalence-test process.

lines E.2 and E.3, successive associations are then made at line A.9. Following this first row association, associations are made recursively and alternately for subsets of columns and subsets of rows (lines E.9 and E.12). The sets S_k and V_k are respectively the set of rows and set of columns of X for which association has been made at the kth level of this recursive process. The *association*( ) procedure is used for making both row *and* column associations. The notation of this procedure is explained by the argument lists and by the formulae at lines E.8 and E.11.

To understand this process one should study carefully the example given in Fig. 7.4. Here we have two matrices with the same row and column invariants which we wish to test for equivalence. A row of type γ or a column of type α has invariant partition (3,2,2, 1,1,0,0) while a row of type δ or a column of type β has partition (3,2,1,1,1,1,0). Thus $Er_0 = \{0,3,4\}$, etc. Our first attempt is to associate row 0 of X with row 0 of Y (at line E.3). This fails when we enter *association*( ) and find that there is no way to associate columns $X_{-,0}$, $X_{-,1}$, and $X_{-,2}$ with $Y_{-,0}$, $Y_{-,1}$, and $Y_{-,2}$ (note that $X_{-,0}$ and $X_{-,2}$ have

A.0 $association(T;"E","Kx","Ky","Lx","Ly")$

A.1   if $Entry = 1: N = \{\ \}$; otherwise: go to $\#\ 200$

A.2   with $t \in T$:

A.3 $\cdot\ \cdot$

A.4   if $\bigvee_{j \in Lx_t \cap Kx}(Q_j \in Ly_{P_t} \cap Ky)$:

A.5 $\cdot\ \cdot$

A.6 $K_t = Lx_t \sim Kx$

A.7   with $k_t \in K_t$:

A.8 $\cdot\ \cdot$

A.9   for $Q_{k_t} \in (E_{k_t} \cap Ly_{P_t}) \sim Ky$:

A.10 $Kx \cup \{k_t\} \rightarrow Kx; N \cup \{k_t\} \rightarrow N$

A.11 $Ky \cup \{Q_{k_t}\} \rightarrow Ky$

A.12 $\cdot\ \cdot$

A.13 $\cdot\ \cdot$

A.14 $Exit = 0$

A.15 $association(\ ) = N$

A.16   exit from procedure

A.17 $\#\ 200$ backtrack

A.18 $Kx \sim \{k_t\} \rightarrow Kx; N \sim \{k_t\} \rightarrow N$

A.19 $Ky \sim \{Q_{k_t}\} \rightarrow Ky$

A.20 $Exit = 1$; exit from procedure

Fig. 7.3. Part II of a *direct* incidence matrix equivalence-test process.

partition α while only $Y_{-,1}$ has partition α). We then drop to line E.16 where that try is undone and return to E.3 for the next possibility, the association of $X_{0,-}$ to $Y_{3,-}$. This does permit the consistent column association $X_{-,0}$ to $Y_{-,4}$, $X_{-,1}$ to $Y_{-,3}$, and $X_{-,2}$ to $Y_{-,5}$. The set $\{0,1,2\}$ then becomes V_1 and we enter *association*( ) again to search for an association for the rows "contained in" those columns. An association is found and we proceed deeper into the nest (lines E.6 through E.15). In this example we do eventually reach the center of the nest where we return to line E.2 to find that we have a complete association—the matrices are equivalent and *equivalence*( ) is set to 1. A summary of the associations and the order in which they are made appear in Fig. 7.4.

The efficiency of this method depends on the degree of segregation of asymmetric rows and columns specified by the Er and Ec sets. For instance, if $Er_i = $ (set) B for each

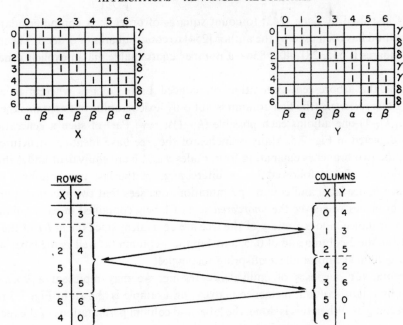

FIG. 7.4. Understanding the equivalence-test process.

$i \in B$, and $Ec_i =$ (set) J for each $i \in J$, the method is very slow-running, while at the other extreme if these sets are singleton sets, there is but a single complete association which must be tested for compatibility. One cannot conclude from this, however, that it is always best to refine the invariant calculation so that asymmetrical rows and columns are most likely segregated *before* the exhaustive search procedure is initiated. This is so because complete segregation is only insured by a complete set of invariants and, short of this, a definite row and column association is still required. The value of such refinement is roughly inversely proportional to the degree of symmetry encountered in the matrix, since extra effort in detecting asymmetry is largely wasted when the configuration is highly symmetric. The invariant calculation and associated Er and Ec sets used here (or their refinement—see exercise 5) represent an effective happy medium in many applications. However, one should experiment before embarking on a long calculation.

7.1.3. *Reduced Latin Square Enumeration*

A *Latin square of order n* is an $n \times n$ 2-array containing the labels $0, 1, \ldots, n-1$ (or any other set of n distinct symbols) in such a way that each symbol appears exactly once in each row and in each column. A Latin square is *reduced* when the elements of the first row and first column are in increasing order. The enumeration of reduced Latin squares is an excellent example of the application of the equivalence algorithm just discussed (and of the branch merging technique of § 4.4.4). The method of enumeration presented

here was first used by Sade [243] to count squares of order 7. A computer adaptation
of this method has been used by the author [254] to count squares of order 8. The number
of such squares is 535, 281, 401, 856—a number unattainable via an explicit generation
scheme.

Imagine the arborescent generation of reduced Latin squares wherein to a $k \times n$
reduced *Latin rectangle* (the first column is not only lexicographic but contains the labels
0, 1, ..., $k-1$) one adjoins each possible $(k+1)$st row. Part of such a generation for
$n = 5$ is depicted in Fig. 7.5. Many branches of the tree have identical structures. For
instance, the two branches emanating from nodes a and b are equivalent under the label
permutation (0,1,2,4,3) followed by an interchange of the last two columns. Besides
equivalence under label and column permutation, one sees that two rectangles produce
identical branches whenever the *unordered* sets of labels contained in each column are
equal. For instance, nodes x and y of the tree are equivalent since the *order* of the labels
0, 2, and 3 in the 2-column and of 0, 3, and 4 in the 4-column (as well as the label sets for
the other columns) cannot affect offspring rectangles.

Therefore, for purposes of equivalence testing, we may represent a $k \times n$ Latin
rectangle by a label-column incidence matrix—an example is shown in Fig. 7.5 for the
x and y rectangles. When this is done, the label and column permutations of the rectangle

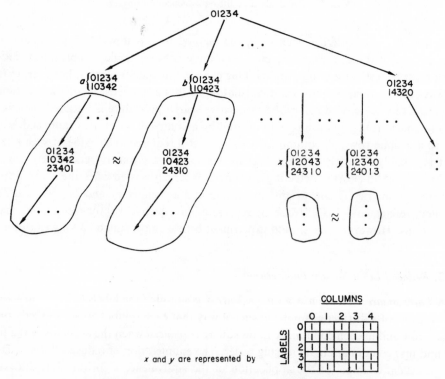

FIG. 7.5. Equivalence of branches in the Latin square generation tree.

become row and column permutations of the incidence matrix and the test for equivalence of two (reduced) Latin squares becomes the test for incidence matrix equivalence discussed in § 7.1.2.

The complete generation for $n = 5$ is illustrated in Fig. 7.6. Each arrow emanating from a rectangle represents one way in which a row may be added to the rectangle. The rectangle to which an arrow points is the representative to which the new rectangle is equivalent. The boxed numbers give the total number of reduced rectangles represented by the corresponding representative rectangle. Since an $(n-1) \times n$ rectangle can be completed to a Latin square in precisely one way [37], 56 is the number of reduced Latin squares of order 5.

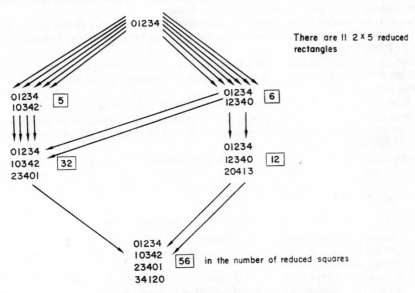

FIG. 7.6. Enumeration of reduced 5×5 Latin squares.

In practice it is sometimes not efficient to carry the branch merging to the bitter end. If, for example, we stopped testing for equivalence after the two 2×5 rectangles were formed, the number 56 would be attained via the computation $(4 \cdot 5) + (2 \cdot 6 + 4 \cdot 6)$ since the rectangle with weight 5 can be completed to a square in four ways and the rectangle with weight 6 can be completed in six ways. In the calculation for $n = 8$, the author found it best to return to the normal tree scan after inequivalent 5×8 rectangles had been calculated via the branch merging procedure.

EXERCISES

1. Write a program which calculates the row and column weights and invariants for an arbitrary incidence matrix.

2. Are the matrices shown in Fig. 7.7 equivalent?

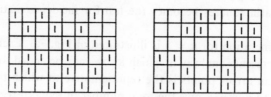

FIG. 7.7. Are they equivalent?

3. Consideration of *reduced* Latin squares in place of *arbitrary* Latin squares amounts to a certain desk precomputation:

(a) Sketch the complete four-level search tree for the generation of Latin squares of order 4 inequivalent under arbitrary permutation of rows and columns.

(b) Devise a modification of the row-by-row generation discussed in this section which effectively incorporates the desk precomputation into the computer program. (Hint: Consider entry-by-entry generation.)

*4. Study the feasibility of enumerating Latin squares of order 9. Consider, among other things, the use of entry-by-entry generation, row-by-row generation, and complete square generation merged into a single program.

5. In studying the example of Fig. 7.4 one notes that $X_{0,-}$ cannot be matched with $Y_{0,-}$ since the associated columns have type (α,α,β) in X and (α,β,β) in Y. Write a program to further segregate asymmetric rows and columns *before* entering *equivalence*( ) by using the set partitions Er and Ec represent to refine themselves.

6. Generalize the programs of Figs. 7.2 and 7.3 so that equivalence is measured relative to a certain *subgroup* of all permutations of rows and columns. In particular consider the subgroup generated by permutations within the individual sets of row labels $\{0,1,\ldots,b-1\}$ $\{b,b+1,\ldots,2b-1\},\ldots,\{(\bar{b}-1)b,\ldots,B-1\}$, where $B = b\bar{b}$, and column labels $\{0,1,\ldots,j-1\},\ldots,\{(\bar{j}-1)j,\ldots,J-1\}$, where $J = j\bar{j}$.

*7. (a) For $n = 3$ through 7, calculate the spectra of determinant magnitudes for the $\binom{2^m}{n}$ distinct $n \times n$ zero–one matrices which do not have duplicate rows. For example, for $n = 2$, three matrices have zero determinant, while three matrices have determinant with a magnitude of one (see exercise 3 of § 4.4).

(b) In exercise 8 of § 6.4 we asked for the number of 5×5 incidence matrices whose permanent is zero. The technique of this section can perhaps be used to enumerate 6×6 matrices with zero permanent. As indicated by part (a), the enumeration of incidence matrices with zero determinant is a slightly more tractable problem. Explain.

*8. For $n = 2$ through 6, calculate $Yn_{m,k}$, the number of distinct $n \times n$ zero–one matrices containing m ones which may be expressed as the union of k different $n \times n$ permutation matrices. For example, $Yn_{n,1} = n!$ and for $n = 3$, $Yn_{6,2} = 6$, and $Yn_{5,2} = 9$.

*9. Write a procedure which will test for the equivalence of two $m \times n$ matrices of real numbers under arbitrary permutation of rows and columns.

7.2. The Steinhaus Sorting Problem

There are many fascinating combinatorial questions which arise in the area of sorting. One such problem, mentioned by Steinhaus [248], is to find the minimum number of binary comparisons necessary to rank n different weights. Let this number be given by $W(n)$. Since there are $n!$ possible weight arrangements and a given comparison can at most discriminate between two equal classes of these possibilities, $W(n) \geqslant L(n)$, where

$$L(2) = 1 \quad \text{and} \quad L(n) = [log_2(n!)]+1 \quad \text{for} \quad n > 2.$$

On the other hand, as pointed out by Steinhaus, use of the binary search procedure (§ 3.2.2) to progressively longer ranked lists shows that $W(n) \leqslant S(n)$, where

$$S(n) = \sum_{m=1 \text{ to } n-1}([\log_2(m)]+1).$$

These bounds for $W(n)$ are equal until $n = 5$. An algorithm of Ford and Johnson [212] provides an improved upper bound $F(n)$ and, in fact, establishes the exact values for $W(n)$ through $n = 11$. A search conducted by this author verifies that $W(12) = F(12)$; this is all that is presently known. A few relevant functional values are given in Table 7.1.

TABLE 7.1

n	$L(n)$	$W(n)$	$F(n)$	$S(n)$
2	1	1	1	1
3	3	3	3	3
4	5	5	5	5
5	7	7	7	8
6	10	10	10	11
7	13	13	13	14
8	16	16	16	17
9	19	19	19	21
10	22	22	22	25
11	26	26	26	29
12	29	30	30	33
13	33		34	37
14	37		38	41
15	41		42	45
16	45		46	50
17	49		50	55
18	53		54	60
19	57		58	65
20	62		62	70
21	66		66	75
22	70		71	80
23	75		76	85
24	80		81	90
25	84		86	95

7.2.1. The Ford–Johnson Algorithm

The binary search procedure is most efficient when the new element is being inserted into a list whose length is one less than a power of two. The Ford–Johnson algorithm takes advantage of this efficiency to a much larger extent than does the Steinhaus algorithm which applies binary search to lists of arbitrary length.

The algorithm is best explained by example: consider the case $n = 15$. The first step is to make 7 ($= [15/2]$) comparisons between fourteen distinct weights; this yields a partial ordering as illustrated in Fig. 7.8(a). The next step is to rank the seven "winners".

This is done with $F(7)$ comparisons by the method being described (thus the algorithm is recursive), yielding a partial ordering as illustrated in Fig. 7.8(b). The third and final step is to place the "losers" (and the odd weight, if it exists) into this ranking of half of the weights. This is done using binary search efficiently as mentioned above. In the example, first w_5 (i.e. weight 5) would be placed in its proper place relative to the list $w_4 > w_2 > w_1$, and then w_3 would be placed relative to w_1, w_2, and w_5. This would give a partial ordering

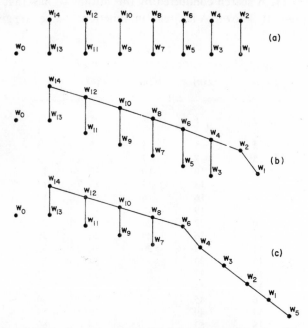

(a)

(b)

(c)

FIG. 7.8. The three steps of the Ford–Johnson ranking algorithm.

(perhaps) as shown in Fig. 7.8(c). Continuing, w_9 would be placed within the known ranking of the seven weights w_1, w_2, w_3, w_4, w_5, w_6, and w_8, and then w_7 would be placed relative to the at most seven weights w_1, w_2, w_3, w_4, w_5, w_6, and w_9. The ranking is completed by inserting w_0 and then w_{13} and w_{11} into the previously sorted lists. In general, during this third step, weights are inserted into a list with length about a power or two minus one.

Analysis of this algorithm yields the recurrence $F(1) = 0$, $F(2) = 1$, and for $n > 2$

$$F(n) = k+$$
$$F(k)+$$
$$\sum_{\substack{m=2,3,\ldots,\text{ as} \\ h=k^*,\ldots,h-z\to h,\ldots,\text{ as} \\ z=2,\ldots,2^{m-1}-z\to z,\ldots,\text{ until } h\leqslant 0,}} [m \times min(h,z)] \qquad (7.1)$$

where $k = [n/2]$ and $k^* = n-k-1$. The first term on the right side of (7.1) comes from the first step of the algorithm, the second term from the second step, and the summation from the third step. The two factors of the summand correspond respectively to the number of comparisons required by a binary search and the number of weight insertions which require that magnitude of search.

As can be seen from Table 7.1, the Ford–Johnson algorithm is very good for small n. However, the difference $F(n) - L(n)$ does increase with n; for example, $F(500) - L(500) = 55$ while $F(1000) - L(1000) = 111$. Thus it is extremely likely that better algorithms exist, although such algorithms are surely more complicated and more difficult to find.

7.2.2. *Enumeration of Poset Consistent Permutations—Discussion*

Before presenting the computer study which establishes that $W(12) = 30$, we discuss an algorithm of central importance in most computer investigations of the Steinhaus problem. This is a means of counting label permutations which describe orderings consistent with a given poset (§ 5.2.4). The significance of this enumeration is that it provides an exact measure of the degree of ordering within the poset. At one extreme an n-poset with $n!$ consistent permutations is unordered (i.e. there are no relations), while at the other extreme an n-poset with but a single consistent permutation is totally ordered (i.e. it is a chain).

The counting procedure described here is based on three easily established facts. Let $M = \{0,1,\ldots,n-1\}$ be the set of labels for an n-poset. For $T \subset M$, denote by $N(T)$ the number of orderings of the labels of T which are consistent with the precedence relations of the poset. For instance, for the poset illustrated in Fig. 7.8(b), $N(\{1,2,3,4\}) = 3$, since the arrangements $(4,3,2,1)$, $(4,2,3,1)$, and $(4,2,1,3)$ describe orderings which are consistent with the relations $w_4 > w_2 > w_1$ and $w_4 > w_3$. The three facts are:

(a) $V \subset M$, $W \subset M$, $V \cap W = 0$, and $\forall_{x \in V; y \in W} (w_x > w_y)$
   $\Rightarrow N(V \cup W) = N(V)N(W)$,

(b) $K1 \subset M$, $K2 \subset M$, $K1 \cap K2 = 0$, and $\forall_{x \in K1; y \in K2} (w_x \not> w_y \text{ and } w_y \not> w_x)$
   $\Rightarrow N(K1 \cup K2) = N(K1)N(K2) \binom{|K1|+|K2|}{|K1|}$, and

(c) $T \subset M$, and $t \in T \Rightarrow N(T) = \sum (N(V)N(W))$, where the summation is over all partitions of $T \sim \{t\}$ into two sets V and W such that $V \cap W = 0$, $V \supset \bigcup_{x \in T \ni w_x > w_t} (\{x\})$, and $W \supset \bigcup_{x \in T \ni w_t > w_x} (\{x\})$.

Statements (a) and (b) follow from the product rule for independent events. The binomial coefficient in (b) gives the number of ways a permutation of the labels in $K1$ and a permutation of the labels in $K2$ can be combined to yield a legitimate permutation of the labels in $K1 \cup K2$. Assertion (c) is a straightforward application of the sum rule for mutually exclusive events and of statement (a).

The idea of the algorithm is to split the problem into simpler and simpler parts, recursively. First, the poset, treated as an undirected graph (i.e. replacing each link by an undirected edge) is tested for connectivity. Each *component* of the problem independently yields an enumeration; these results are eventually combined using (b). To resolve a component, a label t is chosen on which an analysis using statement (c) is fashioned. This analysis produces a set of pairs of simpler problems; each such problem is solved in the recursive manner, the results being combined by the formula of (c).

An example of the process is given in Fig. 7.9: we wish to enumerate permutations for the ten element poset $A_0 = \{2,3,4,5,8,9\}$, $A_1 = \{2,3,4,5,8,9\}$, $A_2 = \{3,4,5,8,9\}$, $A_3 = \{4\}$, $A_4 = \{\ \}$, $A_5 = \{4,8,9\}$, $A_6 = \{4,5,8,9\}$, $A_7 = \{4,5,6,8,9\}$, $A_8 = \{9\}$, $A_9 = \{\ \}$. This poset is shown at the origin of the search tree illustrated in the figure. The poset consists of a single component (shown in the box), so the first step is to resolve that component into

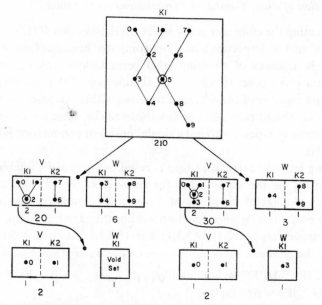

FIG. 7.9. Understanding the recursive procedure for counting poset consistent permutations.

sets of simpler problems per (c). With $t = 5$ (the circled vertex), there are two pairs of problems to consider, i.e. there are two partitions which satisfy the conditions stated in (c). Each pair is shown in two solid boxes at the end of the first level links; the two members of a pair are labeled V and W. Each of these members is then analyzed for connectedness—the components are shown separated by a dotted line and are labeled $K1$, $K2$, ... Each component is then further resolved, the branching process being terminated at either a chain or a void-set since each has $N(\ ) = 1$.

The numbers shown underneath each box of the diagram give the number of permutations consistent with the poset illustrated within the box. They are computed from the formula of (b) when combining components and from the formula of (c) when combining problem pairs. Respectively, for instance, the calculations $2 \cdot 1 \cdot \binom{4+2}{4}$ and $20 \cdot 6 + 30 \cdot 3$ yield the numbers 30 and 210 of the diagram.

7.2.3. *The Enumeration Program*

The enumeration algorithm just discussed may be implemented by the procedures given in Figs. 7.10 and 7.11. The basic recursive process appears in Fig. 7.10 while the auxiliary *component*( ) and problem *pair*( ) generation procedures appear in Fig. 7.11. The answer is the value of the function *measure*$(n, "A", "B")$; $A_{0 \text{ to } n-1}$ and $B_{0 \text{ to } n-1}$ are the incidence matrix and its transpose, respectively, of the given poset.

The principal variables of the recursive process are the 1-arrays of sets V, W and K whose entries give the succession of nested subsets of n which describe the resolved problem pairs (V and W) and the components (K) for a path of the search tree. The initial problem concerns all n vertices, hence $V_0 = \{0, 1, \ldots, n-1\}$. For this problem there may be several components—the possible values for K_1 (in the example of Fig. 7.9 there was but a single component). Each one of these components splits into several pairs of smaller problems given by values of V_1 and W_1, each of these problems has components, and so forth. The nesting terminates when a chain is discovered—when the test of line M.6 is successful.

M.0	$measure(n, "A", "B")$
M.1	$V_0 = (\text{set}) \, n$
M.2	with $i = 1, 2, \ldots$:
M.3	
M.4	$Nw_i = 1$
M.5	for each $component(i; "J", "K")$
M.6	if $\left| \bigcup_{k \in K_i} (\{|B_k \cap K_i|\}) \right| = |K_i|$:
M.7	$S_i = 1$; go to #1
M.8	$S_i = 0$
M.9	for each $pair(i; "V", "W")$:
M.10	$Nv_i = 1$
M.11	
M.12	$V_i = W_i$; $W_i = \{\}$
M.13	if $V_i \neq \{\}$: $Nv_i = Nw_{i+1}$; nest deeper
M.14	$S_i + Nv_i Nw_{i+1} \rightarrow S_i$
M.15 #1	$Nw_i S_i \binom{|J_i| + |K_i|}{|J_i|} \rightarrow Nw_i$
M.16	$measure( ) = Nw_1$; exit from procedure

FIG. 7.10. Main program for counting poset consistent permutations.

The 1-arrays S, Nv, and Nw register various accumulated tallies at each level of the nest. The quantity S_i (formed at line M.14), which ultimately gives the number of permutations for the active component K_i, is the sum of products of fact (c) of § 7.2.2. The quantities Nv_i and Nw_i ultimately give the number of permutations for the pair of problems indicated by V_i and W_i, respectively. These quantities are computed as Nw (line

M.C.0 $component(i; "J", "K")$
M.C.1   if $Entry = 1$: $J_i = \{\ \}$
M.C.2   otherwise: $J_i \cup K_i \rightarrow J_i$
M.C.3 $j = (V_{i-1} \sim J_i)_1$
M.C.4   if nonexistent: $Exit = 1$; exit from procedure
M.C.5 $K, K^* = \{j\}$; $K_i = \{\ \}$
M.C.6   #2   for $k \in \bar{K}$: $K^* \cup [(A_k \cup B_k) \cap V_{i-1}] \rightarrow K^*$
M.C.7   if $K_i \neq K^*$: $\bar{K} = K^* \sim K_i$; $k_i = K^*$; go to #2
M.C.8 $Exit = 0$; exit from procedure

M.P.0 $pair(i; "V", "W")$
M.P.1   if $Entry = 0$:
M.P.2   if $W_i = \{\ \}$: go to #3
M.P.3   otherwise: exit from procedure
M.P.4 $Min = n$
M.P.5   for $k \in K_i$:
M.P.6 $Z = [n \sim (A_k \cup B_k)] \cap K_i$
M.P.7   if $|Z| < Min$: $t_i = k$; $H_i = Z \sim \{t_i\}$; $Min = |Z|$
M.P.8   for $G_i \subset H_i \ni \forall_{g \in G_i}((B_g \cap H_i) \subset G_i)$:
M.P.9 $V_i = (B_{t_i} \cap K_i) \cup G_i$
M.P.10 $W_i = (K_i \sim V_i) \sim \{t_i\}$
M.P.11 $Exit = 0$; exit from procedure
M.P.12   #3   reiterate
M.P.13 $Exit = 1$; exit from procedure

FIG. 7.11. Two auxiliary procedures for the poset-consistent-permutation enumeration program.

M.15) by applying the formula from fact (b) while contributions accumulate from distinct components. The quantity J_i (formed within the *component*( ) procedure) cumulates the vertex labels for the values of K_i already considered.

The *component*( ) procedure generates the components K_i of a given problem M_{i-1} using the precedence relations supplied by A and B. The technique used here (lines C.5 through C.7) is based on the one described in § 6.3.2. The *pair*( ) procedure produces problem pairs V_i and W_i (at lines P.9 and P.10). The vertex on which the "cleavage" [fact (c)] is based is chosen as that vertex for which the number of unrelated vertices (i.e. vertices of the component which neither precede nor succeed the vertex) is a minimum. This vertex t_i and the set of unrelated vertices H_i are computed by the loop at lines P.4 through P.7.

Those vertices of K_i which precede t_i are put in V_i; those which succeed t_i are put in W_i. Each legitimate way in which the vertices of H_i may then be divided between V_i and W_i yields a problem pair. The iteration quantifier of line P.8 generates subsets of H_i which may be included with V_i without contradicting the given precedence relations.

When V_i has been formed (line P.9), W_i becomes the relative (to K_i) complement of this, less t_i (line P.10).

The 1-arrays t, G, and H, subscripted with the nest index i, are used only in this procedure.

To understand this complicated process, the reader should carefully trace the effect of the entire program presented here upon the example discussed in § 7.2.2.

7.2.4. *Computer Study of the Case n = 12*

We now sketch the program which was used to determine that $W(12) > L(12)$, hence that $W(12) = F(12) = 30$.

As a result of a number (say i) of binary comparisons between certain of the n weights, one has a number ($\geqslant i$) of order relationships—these establish a partial ordering among the weights. For such a poset there are several possible pairs of weights for which a relative standing is yet unknown. A comparison between the weights of one of these pairs has two possible outcomes; this yields two poset sons. Thus the entire structure of possibilities may be represented by an arborescence. One part of this tree, a poset and its pairs of poset sons, is illustrated for $n = 4$ in Fig. 7.12. The numbers at the upper left-hand corner of the boxes give the measure for the poset in the box.

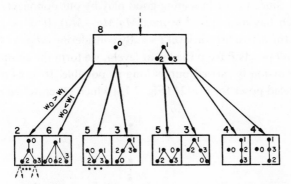

FIG. 7.12. Part of the search tree for a computer study of the Steinhaus problem.

The origin of the tree corresponds to the poset represented by the directed graph with n isolated vertices. The terminal nodes correspond to chains of n vertices. Paths from the origin to a terminal node range in length from $n-1$ to $\binom{n}{2}$ links, corresponding respectively to extremely "lucky" and extremely "stupid" sequences of comparisons. Of course, the number we want is the smallest integer, $W(n)$, that insures that no matter what the actual chain of weights turns out to be we will be able to pinpoint that chain in no more than $W(n)$ comparisons. In this respect it is helpful to view the arborescence as the search tree of variations for a two-person game. For instance, we can imagine that the choice of a pair (x,y) of unrelated weights to be compared next is our option, while the outcome of the comparison (e.g. $w_x > w_y$ or $w_y > w_x$) is our opponent's option. The problem is then to determine, assuming best play by our opponent, the fewest number of "full" moves (a move by both players) needed to reach a terminal node.

Because of the large size of this tree for $n = 12$, an exhaustive scan is out of the question. However, it *is* feasible, using equivalences between labeled posets and reasonable impasse detection, to scan the tree for the *possible* existence of a strategy having $W(n) = k$. If, as in the case of $n = 12$ and $k = 29$, the scan shows that no strategy can exist, then $W(n) > k$. The idea is to show that our opponent can always make moves good enough to prevent us from reaching a conclusion in 29 moves regardless of our strategy. This is accomplished using the tier scan approach of § 4.4.1 (see especially Fig. 4.23).

More specifically, consider the following recursion: Start at level 0 with the unique unlabeled poset consisting of twelve isolated vertices. This poset has measure 12! which is less than 2^{29}. Now imagine that at level i we have an "active" list of unlabeled posets each of which arose from i comparisons and has measure less than 2^{29-i}. From these posets we derive a set of posets at level $i+1$. Let $P_{0 \text{ to } 11}$ be a poset at level i with measure Mp. Let x and y be two unrelated elements of P. Initialize Q and Q^* to P and then let $Q_x = P_x \cup \{y\}$ and $Q_y^* = P_y \cup \{x\}$; that is, Q is the poset derived from P by assuming that $w_x > w_y$, and Q^* from P by assuming that $w_x < w_y$. Let Mq be the measure of Q as calculated by the *measure*( ) procedure discussed earlier; by the rule-of-sum the measure of Q^* is $Mp - Mq$. Since we are assuming good play by our opponent, we consider only Q or Q^*, one which has measure $M = max(Mq, Mp - Mq)$. If $M \leqslant 2^{29-(i+1)}$ then this poset is placed on the active list for level $i+1$. By considering *all* pairs $\{x,y\}$ of unrelated vertices for P and *all* posets P from the list at level i, we form the complete active list for level $i+1$. This recursion is carried out as long as possible. It turns out that there is a single active unlabeled poset for $i = 24$ (Fig. 7.13) which generates *no* posets for $i = 25$. Thus $W(12) > 29$.

FIG. 7.13. The last surviving poset.

In this generation the labeling of the elements of the poset serves only to permit representation of the poset within the computer; the structure of the unlabeled poset contains the relevant information. Thus only representative unlabeled posets need be saved at each level of the generation. Since the same unlabeled poset can arise during the scan from quite different labeled posets, this aspect of the generation involves equivalence testing between posets. The posets are simply directed graphs, hence the method of § 7.1 can be applied (but see exercise 4). Note also that as far as the Steinhaus problem is concerned a poset may be "inverted" (the order relation may be reversed) without altering the structure of the poset.

EXERCISES

*1. Write a program which implements the Ford–Johnson sorting algorithm. Calculate the exact average number of comparisons required by that algorithm for $n = 2$ through 8. For $n = 9$ through 15 calculate an approximate average by considering only a random sample of permutations.

*2. Note from Table 7.1 that the function $F(n) - L(n)$ is not a monotonically increasing function of n. What does this suggest in the way of an improved sorting algorithm?

*3. The chain-test at line M.6 of Fig. 7.10 is not an essential part of the program; it is included for efficiency.
   (a) Simplify the program of Fig. 7.10 by eliminating this test, letting the recursive process continue until all vertices are isolated.
   (b) Study the effect upon the efficiency of the algorithm created by including *more* elaborate tests at line M.6, e.g. a test for a V-shaped poset.

4. Since poset equivalence testing is used in the § 7.2.4 study merely to reduce the size of the scan, a definitive equivalence test is not essential to the algorithm. Use of the high-powered complete method of § 7.1 is perhaps not necessary in such studies. Devise a simpler *nonequivalence* test which can be effectively used in studies of this sort.

*5. It happens that the poset of Fig. 7.13 does *not* result from a strategy of play which says that we should always make a move (i.e. choose a comparison) which maximizes the smaller measure of the two poset sons. However, such a strategy could conceivably reach a terminal node in thirty moves. Investigate these "reasonable" strategies for $n = 12$ and $n = 13$.

6. The recursive computations of the three simultaneously running indices of the summation of (7.1) are written as subscripts to indicate an efficient means of calculation. (The display used there points up a possible, not unnatural extension of our combinatorial language.) Rewrite this summation with a single index m and with z and h of the summand replaced by functions of m.

7. Show that the poset of Fig. 7.13 cannot be ordered in five more comparison-steps.

7.3. A Computer Study of the Four-color Problem

The *four-color problem* is perhaps the most famous unsolved combinatorial problem. We outline here one approach for investigating this fascinating puzzle. We use the vertex coloring formulation broached in § 4.2. The question is: Can the vertices of every maximal planar graph (see exercise 7 of § 4.2) be colored with four colors in such a way that adjacent vertices have distinct colors?

7.3.1. *Elementary Kempe Reductions*

A line of attack on the four-color problem which has received much attention is the search for *reducible configurations*, i.e. subgraphs which could not exist in the "smallest" uncolorable maximal planar graph (if it exists). A vertex with valence three is a simple example of a reducible configuration. Suppose, for instance, that such a vertex, v_3, did exist in a given graph G. Since G is maximal, v_3 would appear in the graph as shown in Fig. 7.14(a). Consider the graph G^* formed from G by the elimination of v_3 and the edges which contain it. If G^* is colorable (Fig. 7.14(b)), then so is G (Fig. 7.14(c)) since at most three colors are required for v_0, v_1, and v_2 and the fourth is available for v_3. (We use the labels α, β, γ, δ for the four colors. Of course, numeral labels, e.g., 1, 2, 3, 4, are used within the computer.) It is therefore sufficient to consider the smaller graph G^*, or in general to consider only graphs without valence three vertices, in any search for an uncolorable graph.

It is only slightly more difficult to show that a graph G containing a vertex of valence four is also reducible. Consider the subgraph H embedded in G—Fig. 7.15(a). Since G is planar the edges $\{v_0,v_2\}$ and $\{v_1,v_3\}$ cannot both exist in G. Assume $\{v_1,v_3\}$ is not in G. We may form a smaller graph G^* identical to G external to H but with vertices v_1, v_3 and v_4 "contracted" to a single vertex u—Fig. 7.15(b). In this contraction the edges $\{v_1,v_4\}$, $\{v_3,v_4\}$ and $\{v_0,v_1\}$, $\{v_1,v_2\}$, etc., have been eliminated. If G^* is colorable, at most three colors are needed for v_0, v_2, and u. As before, we can then adopt this "coloration" (i.e. consistent coloring) for G, extending it to v_4 as shown in Fig. 7.15(c).

There is an alternate proof of this reduction. Suppose that instead of contracting along the edges $\{v_1,v_4\}$ and $\{v_3,v_4\}$ (still under the assumption that $\{v_1,v_3\}$ is not in G), we eliminate vertex v_4 and its edges and adjoin edge $\{v_1,v_3\}$ as shown in Fig. 7.16(b). Colorations

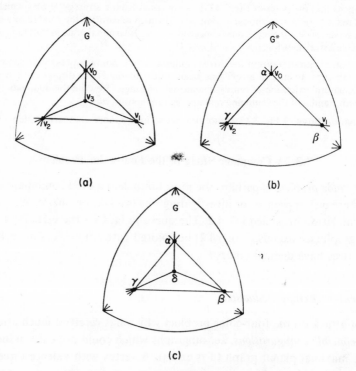

FIG. 7.14. Proving the reducibility of a 3-vertex.

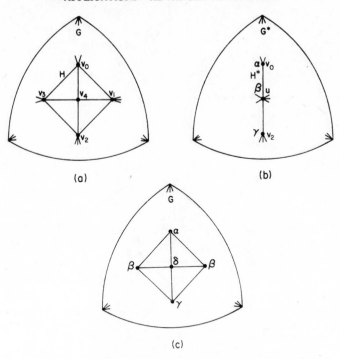

FIG. 7.15. Proving the reducibility of a 4-vertex, method 1.

of G^* (if they exist) are, up to isomorphism, of two types: $(\alpha,\beta,\alpha,\gamma)$, that is, $v_0 = \alpha$, $v_1 = \beta$, $v_2 = \alpha$, and $v_3 = \gamma$ as shown in Fig. 7.16(b) or $(\alpha,\beta,\gamma,\delta)$ as shown in Fig. 7.16(c). Colorations of the first type can be extended to G immediately since δ is available for v_4. Consider a coloration of the second type; we show how this coloration can be transformed so that only three colors appear on the *circuit* (v_0,v_1,v_2,v_3).

Let $G\alpha\gamma$ be the "section" subgraph (see exercise 3 of § 3.4) of G^* whose vertices are those colored α or γ. We call such a subgraph an (α,γ)-Kempe subgraph. Since any vertex in G^* adjacent to a vertex of $G\alpha\gamma$ but not in $G\alpha\gamma$ is colored β or δ, the colors α and γ may be interchanged on all the vertices of any component of $G\alpha\gamma$ without destroying the consistency of the coloring scheme. Similarly, let $G\beta\delta$ be the Kempe subgraph defined by vertices colored with a β or δ; the colors on the vertices of a component of this subgraph may also be interchanged. Consider the component, $K0$, of $G\alpha\gamma$ which contains v_0. If $K0$ does not contain v_2, then an interchange of colors on the vertices of $K0$ produces a coloration of G^* with $(\gamma,\beta,\gamma,\delta)$ on the circuit (v_0,v_1,v_2,v_3) so that α may be used for v_4 in the extension to G. If $K0$ does include v_2, then there is a path from v_0 to v_2 with vertices of the path alternately having colors α and γ—such a path is called an (α,γ)-Kempe path from v_0 to v_2. In this case, consider the component, $K1$, of $G\beta\delta$ which contains v_1. If $K1$ contained v_3, then there would be a (β,δ)-Kempe path from v_1 to v_3. This is impossible by the planarity of G^* (since it must cross the (α,γ)-Kempe path), so we conclude

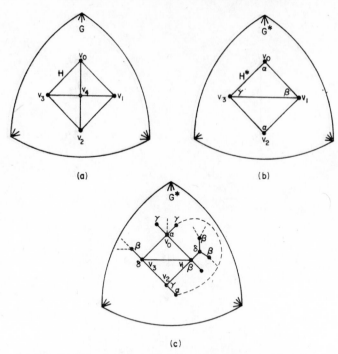

FIG. 7.16. Proving the reducibility of a 4-vertex, method 2.

that $K1$ does not contain v_3. But then the colors on the vertices of $K1$ can be interchanged to yield $(\alpha,\delta,\gamma,\delta)$ on our circuit, and vertex v_4 can receive color β.

In 1879, Kempe [225] mistakenly thought that a vertex with valence five was also a reducible configuration. If this were true the four-color problem would be solved since it is a simple consequence of Euler's formula (i.e. $V-E+F = 2$) that a maximal planar graph with no 3-vertices or 4-vertices contains a 5-vertex (in fact, at least 12 5-vertices—see exercise 1 and [235]). Kempe's error was discovered by Heawood [222] in 1890. Since that time (correct) generalizations of Kempe's original argument have been used by various authors (notably, Franklin [213], Birkhoff [201], Winn [256], and Bernhart [199]—see Ore [235], chapter 12, for a comprehensive account) to establish the reducibility of assorted configurations. Unfortunately, the reducible configurations known to date (see Appendix III) are too unrestrictive to solve the problem.

7.3.2. *A General Approach to Reducibility Testing*

Consider a connected planar graph consisting of either an isolated vertex or of triangles and isthmic and peninsular edges as illustrated in Fig. 7.17. In this discussion we call such a graph a *configuration*. Let I be a configuration having a potential valence associated with each "outside" vertex. One examines the ways in which I can be contained as a section subgraph in a maximal planar graph having the valence of each vertex of I

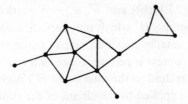

FIG. 7.17. Example of a "configuration".

equal to its potential valence. If in all cases there exists a maximal planar graph G^*, having fewer vertices than G, whose colorability implies the colorability of G, then I is a *reducible configuration*.

In this book we consider only those configurations which can be embedded within a maximal planar graph in a unique way. Furthermore, for a given configuration I, we assume that the vertices in $G \sim I$ which are connected to a vertex of I form a circuit in G. Thus, by maximality, there is a well-defined subgraph H, called the *environment* of I, formed by adjoining this circuit to the "star" of I (see exercise 4 of § 3.4). For our purposes we further assume that H is a section subgraph, i.e. that the only connections

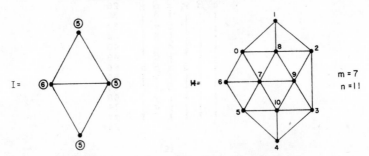

FIG. 7.18. A sample configuration and environment.

in G between circuit vertices are circuit edges. (Due to existing theorems—e.g. Th. 12.1.5 of Ore [235]—these assumptions are plausible for relatively small configurations and environments and do allow meaningful results to be achieved.)

An example of a configuration I and its environment H appears in Fig. 7.18. The circled numbers on the vertices of I are the potential valences. The vertices of H are labeled $0, 1, \ldots, m-1, m, m+1, \ldots, n-1$ where the first m labels are assigned sequentially to the *circuit* vertices and the final $n-m$ labels are assigned arbitrarily to the *interior* vertices (i.e. the vertices of I).

Associated with a given configuration environment is an incidence system consisting of "circuit colorations" as the objects and "contractions" as the blocks. A *circuit coloration* (more precisely a labeled-circuit coloration) is a vector $c_{0 \text{ to } m-1}$ describing a coloration of the circuit. A circuit coloration is normalized when $c_0 = \alpha$ and $c_1 = \beta$. A *contraction* is any configuration which has m "circuit vertices" (m not necessarily distinct encircling vertices—see Fig. 7.19(b)) and n^*, $n^* < n$, vertices in all. In other words, a contraction is simply a subgraph H^* which may replace a configuration environment H in a graph G to produce a smaller maximal planar graph G^*. (The employment of the word "contraction" in this context is perhaps more historical than graphic. Note that the interior of H^* need not be related to the interior of H.) A contraction *contains* a circuit coloration if that coloration applied to the circuit of the contraction can be extended to a consistent coloring of the entire contraction. Figure 7.19 illustrates (a) an environment,

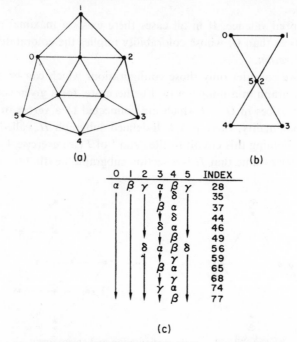

(c)

FIG. 7.19. The circuit colorations of a sample contraction.

(b) a contraction, and (c) the normalized circuit colorations which the contraction contains.

With $\alpha \equiv 1$, $\beta \equiv 2$, $\gamma \equiv 3$, $\delta \equiv 4$, a convenient serial number representation for a normalized circuit coloration (essentially a conversion from base 3 notation) is the index

$$\sum_{i=2 \text{ to } m-1} [|(\{1,2,3,4\} \sim \{c_{i-1}\}) \cap c_i| \times 3^{m-1-i}]. \tag{7.2}$$

This index does not exceed 3^{m-2} for a circuit with m vertices, hence one can work within a universe of $\{0,1,\ldots,3^{m-2}-1\}$. A contraction is therefore suitably represented by the set of indices corresponding to the circuit colorations it contains—Fig. 7.19(c).

The general process of testing the reducibility of a given configuration is to search for a contraction which contains only "tractable" circuit colorations, namely those that may be consistently extended into the interior of H or that may be transformed by means of Kempe interchanges into such extendable colorations. If such a contraction is found, the configuration is reducible since the uncolorability of any graph containing the configuration implies the uncolorability of a smaller graph.

In the first proof of the reducibility of a 4-vertex (Fig. 7.15) a contraction was found which contains, up to isomorphism, only a single circuit coloration. This coloration is "zeroth-order tractable", it is extendable without considering Kempe paths. In the second proof (Fig. 7.16), the discovered contraction contains two circuit colorations; one is zeroth-order tractable and the other, requiring Kempe interchanges, is "first-order tractable". For computer work it is best first to calculate the tractability of all circuit colorations and then to search for a contraction which contains only tractable circuit colorations.

7.3.3. *Tractable Circuit Colorations*

Let H be a given configuration environment for which we wish to calculate the tractable circuit colorations. Since a permutation of colors or a dihedral transformation leaving the environment unchanged takes one coloration of the entire environment into another coloration, it is sufficient to consider circuit colorations inequivalent under these permutations and transformations. Let S be the index set for a set of such representative circuit colorations. By (7.2), $S \subset$ (set) 3^{m-2}. The first step is to calculate, by the techniques of § 4.2 perhaps, those circuit colorations which are zeroth-order tractable, i.e. those which are, without alteration, part of a consistent coloring of the environment. We denote this subset of S by T_0; the zeroth-order *intractable* circuit colorations are indicated by $I_0 (= S \sim T_0)$. By a method to be discussed shortly one next calculates the set of first-order tractable circuit colorations T_1 and the set $I_1 = I_0 \sim T_1$. In the same way the set T_2 and the set $I_2 = I_1 \sim T_2$ are calculated. This process is repeated until no new tractable circuit colorations arise, i.e. until $T_k = \{ \ \}$. The set $R = S \sim I_{k-1}$ $[= \bigcup_{i \in k}(T_i)]$ gives the tractable circuit colorations. If $R = S$, as is many times the case, any contraction contains only tractable colorations and the configuration is *a priori* reducible.

To explain the missing link of this recursive process, we now change the notation slightly. Let T, $T \subset S$, indicate a set of circuit colorations known to be tractable. Let c, $c \in S \sim T$, be a circuit coloration which we wish to test for tractability. As decribed in § 7.3.1, a coloration of a graph G may be transformed by interchanging color x with color y on the vertices of a component of the section subgraph Gxy which contains all vertices colored with an x or with a y; $x, y \in \{\alpha, \beta, \gamma, \delta\} \ (\equiv \{1,2,3,4\})$ and $x < y$. The effect of this interchange on c depends, of course, upon the intersection of c with that component—all vertices of c contained in the component have their color changed. If the transformed circuit coloration is tractable, then c itself is tractable with respect to all graphs which allow colorations having a Kempe component which has this same intersection with c. If for every possible component structure there exists a set of legitimate color pair interchanges which produces a tractable circuit coloration, then c is itself tractable.

In computer language the complete test can be written

$$\text{if } \exists_{split(\ )} [\forall_{structure(\ )} (\exists_{interchange(\ )} [transform(c) \in T])] : \tag{7.3}$$

where *split*( ), *structure*( ), and *interchange*( ) are appropriately defined generation procedures and *transform*(c) yields the circuit coloration derived from c by the indicated color interchange. Of course, a color interchange may not produce an element of S since S consists of equivalence class *representatives*. Thus *transform*( ) must contain a program to convert the resulting circuit coloration to the canonical form chosen for elements of S.

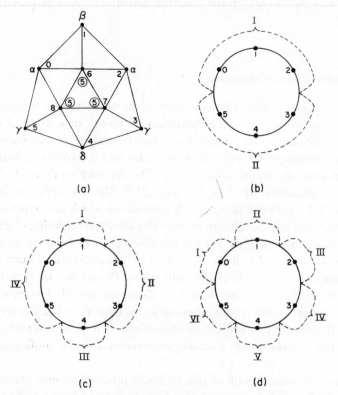

FIG. 7.20. Example of the intersection of the possible Kempe subgraphs with the circuit of an environment.

The procedure *split*( ) considers the three types of Kempe subgraph decompositions $G\alpha\beta$ and $G\gamma\delta$, $G\alpha\gamma$ and $G\beta\delta$, and $G\alpha\delta$ and $G\beta\gamma$. For each of these decompositions we examine the possible component *structures*—more specifically, the possible ways in which component structures can intersect with the circuit coloration. For each structure we hope to find an *interchange* which allows a transformation to a tractable circuit coloration.

As an example, consider the configuration environment and circuit coloration given in Fig. 7.20(a). This circuit coloration is not zeroth-order tractable since vertices 7 and 8 must both receive the color β. It *is* first-order tractable, however, as can be seen by the following analysis.

The three types of Kempe subgraph decomposition are illustrated in (b), (c), and (d) of Fig. 7.20. Fix your attention on the subgraphs $G\alpha\delta$ and $G\beta\gamma$, that is, on (d) of the figure. The possible component structures may be expressed by partitions of $\{I,II,III, IV,V,VI\}$. (We temporarily use Roman numerals to label blocks of adjacent circuit vertices which serve as elements of the set being partitioned. This is to avoid confusion with and emphasize their difference from vertex labels.) For instance, the structure $\{I\}$ $\{II,IV\}$ $\{III\}$ $\{V\}$ $\{VI\}$ has vertices 1 and 3 in the same component (of $G\beta\gamma$, of course) and vertices 0, 2, 4, and 5 each in separate components. Actually, we need only consider structures corresponding to *maximal* partitions (see § 5.5.4) since an interchange which works for a maximal partition also works for a refinement of that partition.

There are five (a Segner number!) basic component structures:

(1) $\{I\}$ $\{II,IV,VI\}$ $\{III\}$ $\{V\}$,   (2) $\{I,V\}$ $\{II,IV\}$ $\{III\}$ $\{VI\}$,   (3) $\{I\}$ $\{II,VI\}$ $\{III,V\}$ $\{IV\}$,

(4) $\{I,III\}$ $\{II\}$ $\{IV,VI\}$ $\{V\}$,   (5) $\{I,III,V\}$ $\{II\}$ $\{IV\}$ $\{VI\}$.

Each of these structures permits a number of color interchanges. For instance, structure (1) permits interchange of the color (with its mate) on the vertices of block I (here, vertex 0); the vertices of blocks II, IV, and VI (here, vertices 1, 3, and 5); the vertices of block III (here, vertex 2); the vertices of block V (here, vertex 4); and on the vertices of block I *and* blocks II, IV, and VI; etc. Here we see that an interchange on vertex 4 (i.e. replacing δ by α) yields a tractable circuit coloration since vertices 6, 7, and 8 can then be colored γ, β, and δ, respectively. Similarly, each of the other four structures permits an interchange transformation to a tractable coloration (the reader should verify this), hence our original coloration is tractable.

In practice it is advisable to precompute an interchange-structure incidence matrix for each possible partition size. When the circuit of the configuration environment contains m vertices, the splits can produce partitions of $\{0,1,\ldots,2j-1\}$ for $j = 1$ to m. (We now use Arabic numerals to label the elements of the set being partitioned.) The partition of $\{0\}$ can be ignored. The matrix for $j = 3$ is given in Fig. 7.21. Note that the labels of the set being partitioned correspond to the vertex labels only when $j = m$ as in (d) of Fig. 7.20 (i.e. I $\equiv$ 0, II $\equiv$ 1, ..., VI $\equiv$ 5). When $j < m$, some of the elements of the set being partitioned correspond to more than one circuit vertex—see Fig. 7.20(b) and (c).

We leave the programming details of tractable circuit coloration calculation to the reader—see the exercises. Actually the programs discussed here either appear explicitly elsewhere or are closely related to algorithms discussed elsewhere in this book. For instance, the calculation of the structure-partitions appears explicitly in Fig. 5.30 and the convergence to the set of tractable colorations resembles the component calculation of Fig. 6.12. Also, with the interchange-structure matrices available, the implementation of (7.3) is actually a series of quite elementary covering problems.

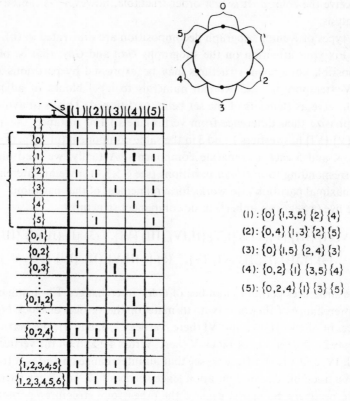

FIG. 7.21. The interchange-structure incidence matrix for $j = 3$.

7.3.4. *Contractions*

Having calculated the set T of tractable circuit colorations of a given configuration environment, the next step in testing for reducibility of the configuration is to search for a contraction which contains only elements of T. In computer language, the configuration is reducible when the test

$$\text{if } \exists_{contraction(H) \to C} \left[\forall_{ccoloration(c) \to c} (c \in T) \right]: \qquad (7.4)$$

is successful. Here, *contraction*( ) is a generation procedure which produces in turn each contraction C of the configuration environment H, and *ccoloration*( ) is a generation procedure which produces all (canonical form) circuit colorations of this contraction.

Contractions of a given configuration environment are configurations used as trial replacements for the environment (in order to produce smaller, hence hypothetically colorable, graphs). Thus contraction generation is configuration (as defined in § 7.3.2) generation. The generation of *all* contractions is a formidable problem in its own right. We content ourselves here with a few general comments. Actually many reducible configurations can be discovered by considering a subclass of contractions (see exercise 9). Such a method was in fact used by the author to construct Appendix III.

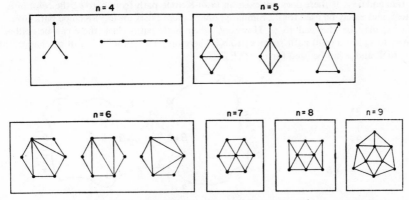

FIG. 7.22. Contractions with less than ten vertices.

The configurations with six vertices on the circuit ($m = 6$) and fewer than ten vertices in all ($n < 10$) to be used as contractions are illustrated in Fig. 7.22. Reducible configurations, e.g.

need not be used as contractions since it is sufficient to use the reduced version

when searching for an effective contraction. In general, each configuration with m circuit vertices may be inserted as a contraction in $2m$ ways, one way for each member of the dihedral group. Of course, symmetries of the configuration can be used to reduce this number.

There is much to be learned about contractions and their effectiveness in establishing the reducibility of configurations. While "small" contractions (e.g. the first two of Fig. 7.22) are often better since they contain fewer circuit colorations, one cannot *a priori* disregard "large" contractions (e.g. the last two of Fig. 7.22) in any particular problem.

Indeed, when the four-color problem is solved, it is extremely likely that elaborate computer investigations will have played a nonnegligible role.

EXERCISES

1. (a) Prove that every maximal planar graph with no 3- or 4-vertices contains at least twelve -5 vertices. (Hint: Eliminate F from Euler's formula $V - E + F = 2$.)
   (b) Prove that every maximal planar graph can be consistently colored using five colors.

2. What is wrong with the following (Kempe's) "proof" of the four-color conjecture? Consider the contraction given in Fig. 7.23(a) for the environment of a vertex with valence five which appears in Fig. 7.23(b). The only circuit coloration which is not immediately extendable to the interior is as shown. Consider $G\alpha\delta$ and $G\alpha\gamma$. If there does not exist an (α,δ)-Kempe path from v_0 to v_2, the color of v_0 may be changed to δ, and α may be used for v_5. Similarly, unless there exists an (α,γ)-Kempe path from v_0 to v_3, the color of v_0 may be changed to γ. However, when both paths exist, there can be neither a (β,γ) path from v_1 to v_3 nor a (β,δ) path from v_4 to v_2. Therefore, the color of v_1 can be changed to γ, the color of v_4 to δ, and β can be used for v_5. (Q.E.D.!)

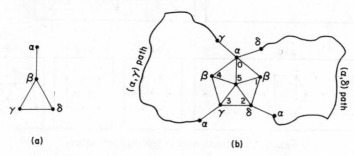

(a)   (b)

FIG. 7.23. Kempe's "proof" of the four-color conjecture.

3. We have assumed in this section that the configuration environment Z is a *section* subgraph bounded by a *circuit*. Discuss the modification of the algorithm necessary when circuit vertices are not actually distinct and when particular "external" edge connections between circuit vertices are allowed.

*4. (a) Write a program which calculates the circuit coloration corresponding to an index N, $N \in 3^{m-2}$. Note that (7.2) gives the inverse calculation.

(b) The actual number of labeled-circuit colorations distinct under color permutations is

$$\left[\frac{3^{m-1}+5}{8'} \right],$$

where m is the number of circuit vertices (or is given by the recursion $c_1 = 0$, $c_2 = 1$, $c_m = 2c_{m-1} + 3c_{m-2} - 1$ for $m \geqslant 3$). Devise a serial number indexing scheme for circuit colorations more efficient than suggested by (7.2).

5. Write a program which calculates the set S, $S \subset 3^{m-2}$, which indicates the zeroth-order tractable circuit colorations for a given configuration environment.

6. Verify in detail the tractability of the circuit coloration of Fig. 7.20(a).

*7. Note in Fig. 7.21 that the interchange involving vertex blocks 0 and 2 is the same as the interchange involving block 4. This is so because blocks 0, 2, and 4 are associated with components of the same Kempe subgraph and interchanging the colors on vertices of this subgraph does not affect the component structure. There are effectively nine distinct interchange possibilities for $j = 3$, the eight bracketed ones of Fig. 7.21 plus $\{2,5.\}$.

(a) Construct the interchange-structure matrix for $j = 4$.

(b) How many operative interchange possibilities are there in general?

8. Explain why it is only necessary to use configurations and not general "non-maximal" subgraphs, e.g.

as contractions.

9. Adapt the program of exercise 6 of § 3.5 for generation of a subclass of contractions.

10. Show that the configuration consisting of a triangle of vertices each with potential valence five (sometimes called a *triad*) is not reducible by this method.

11. Draw environments for each of the configurations given in Appendix III.

*12. Find a new reducible configuration.

7.4. Parker's Orthogonal Latin Square Generation

We present here a general computer method for generating Latin squares orthogonal (see exercise 11 of § 5.1) to a given Latin square. The method is due to Parker [236] who has successfully applied it to the case of 10×10 squares. This method was developed *after* Euler's conjecture that no 10×10 orthogonal squares did exist was refuted by Parker [237] and by Bose, Shrikhande, and Parker [203].

Let $L_{0 \text{ to } n-1, 0 \text{ to } n-1}$ be a Latin square of order n; that is, $L_{i,j} \in n$, $\bigcup_{x \in n} (\{L_{i,x}\}) = n$, and $\bigcup_{x \in n} (\{L_{x,j}\}) = n$, for all $i \in n$ and $j \in n$. A *transversal* of L is a permutation $(t_0, t_1, \ldots, t_{n-1})$ of $(0, 1, \ldots, n-1)$ such that

$$\bigcup_{x \in n} (\{L_{x,t_x}\}) = \{0, 1, \ldots, n-1\} \tag{7.5}$$

is true. Two transversals $t_{0 \text{ to } n-1}$ and $v_{0 \text{ to } n-1}$ are *disjoint* when $\bigvee_{i \in n} (t_i \neq v_i)$ is true. A set of n mutually disjoint transversals $T_{0,-}, T_{1,-}, \ldots, T_{n-1,-}$, yields a Latin square $M_{0 \text{ to } n-1, 0 \text{ to } n-1}$ orthogonal to L defined by

$$M_{i,j} = k \Leftrightarrow T_{k,i} = j. \tag{7.6}$$

To understand this, note first that M is indeed a Latin square: one, each row contains each label exactly once since the transversals are disjoint, and, two, each column contains each label exactly once since each transversal by (7.5) covers all columns and by (7.6) determines the placement in M of a *particular* label. Now for M to be orthogonal to L, the n^2 pairs $(L_{i,j}, M_{i,j})$ must be distinct. Consider the n entries of M which are equal to k. By (7.6) these are determined by the transversal $T_{k,-}$ of L. The transversal definition (7.5) insures that the entries

$$L_{x,T_{k,x}}, \ x \in n$$

(which are associated with these entries of M), are distinct, hence the n pairs with k as a second component are distinct. By considering the n distinct possible values for k, we see that indeed n^2 distinct pairs do exist and the squares are orthogonal. An example of this construction for $n = 5$ appears in Fig. 7.24 (the circled entries of L indicate the $T_{0,-}$ transversal).

	0	1	2	3	4
0	⓪	1	2	3	4
1	1	4	0	②	3
2	2	0	3	4	①
3	3	2	④	1	0
4	4	③	1	0	2

L

$T_{0,-} = (0,3,4,2,1)$

$T_{1,-} = (1,2,3,0,4)$

$T_{2,-} = (2,4,1,3,0)$

$T_{3,-} = (3,1,0,4,2)$

$T_{4,-} = (4,0,2,1,3)$

0	1	2	3	4
4	3	1	0	2
3	2	4	1	0
1	4	0	2	3
2	0	3	4	1

M

FIG. 7.24. An example of Parker's orthogonal Latin square formation.

T.1 $S_0 = $ (set) $n;\ D_0 = \{\ \};\ N = 0$
T.2   with $i = 0$ to $n-1$:
T.3 $\cdot\ \cdot$
T.4   for $t_i \in S_i \ni L_{i,t_i} \notin D_i$:
T.5 $S_i \sim \{t_i\} \to S_{i+1}$
T.6 $D_i \cup \{L_{i,t_i}\} \to D_{i+1}$
T.7 $\cdot\ \cdot$
T.8 $T_{N,0 \text{ to } n-1} = t_{0 \text{ to } n-1};\ N+1 \to N$
T.9   "$T_{0 \text{ to } N-1,-}$ are the N transversals"
I.1   for $i = 0$ to $N-1$:
I.2 $B_i = N \sim (i+1)$
I.3   for $j = i+1$ to $N-1$:
I.4   if $\exists_{k \in n}(T_{i,k} = T_{j,k})$: $B_i \sim \{j\} \to B_i$
I.5   "B_i gives those transversals which are disjoint from ith transversal"

G.1 $A_0 = $ (set) N
G.2   with $k = 0$ to $n-1$:
G.3 $\cdot\ \cdot$
G.4   for $a_k \in A_k$
G.5 $A_{k+1} = A_k \cap B_{a_k}$
G.6 $\cdot\ \cdot$
G.7   COMPUTE AND USE THE ORTHOGONAL
  $\vdots$ MATE M FROM $T_{a_0,-}, T_{a_1,-}, \ldots, T_{a_{n-1},-}$

FIG. 7.25. A program to search for an orthogonal mate.

We see from this discussion that the generation of squares orthogonal to a given L is a two-step process: (1) generate transversals of L, and (2) generate n-sets of distinct transversals. The transversal generation is simply a permutation generation with impasse detection to insure that (7.5) is satisfied, while the n-set generation is a disjoint covering problem. The complete generation is sketched in Fig. 7.25 with segment T giving step (1) and segment G giving step (2). The calculation which must be performed to prepare the set of transversals for the covering algorithm of segment G appears as segment I.

Actually, due to the large number of transversals for an arbitrary interesting (e.g. $n \geqslant 10$) Latin square, the straightforward precomputation and covering algorithm shown here is impractical—e.g. when $N = 1000$, about 5×10^5 executions of line I.4 are required. A reasonable compromise consists of precomputing (perhaps in segment T)

$$C_{i,j} = \bigcup_{x \in N \ni T_{x,i} = j}(\{x\}), \tag{7.7}$$

for $i \in n$ and $j \in n$, and replacing line G.5 by

$$\text{G.5} \qquad A_{k+1} = A_k \sim \left[\bigcup_{i \in n}(C_{i,T_{a_k,i}}) \cup (a_k+1) \right] \tag{7.8}$$

for instance. The mask $C_{i,j}$ gives the indices of transversals which have label j in the ith position. These masks are then used in segment **G** to *calculate* the set of transversals which are not disjoint from $T_{a_k,-}$ (or which precede $T_{a_k,-}$). This set is the complement of B_{a_k}, hence subtraction is used in line G.5 instead of intersection.

EXERCISES

1. In Fig. 7.24, the T's themselves form a Latin square orthogonal to L. Construct a Latin square which contains an orthogonal mate whose corresponding transversals do not form a square orthogonal to the original square.

2. Write the program indicated in Fig. 7.25 by G.7 and following lines.

***3.** Find a pair of orthogonal 10×10 Latin squares.

7.5. Simplification of Normal Form Boolean Expressions

Consider the problem of finding "minimal" normal form expressions for a given Boolean function. Recall from § 3.5 that a normal form expression is a sum of products of "literals" (primed or unprimed variables). Such an expression is *minimal* if it contains as few occurrences of literals as possible. For example, the expression

$$x'y' + yz + xz' ,$$

which contains six literal occurrences, is a minimal form for the three variable function $\{0,1,3,4,6,7\}$; the expression

$$x'y' + x'z + xy + xz' , \tag{7.9}$$

which contains eight literal occurrences, though irredundant, is nonminimal.

Let f be a Boolean function of n variables given as a subset of vertices of the n-cube for which we wish to calculate a minimal normal form expression. For practical reasons we assume $n \leqslant 9$. Since a prime implicant of the function has fewer literal occurrences than any cell which it contains (i.e. it has $n - d$ literal occurrences where d is the dimension of the prime implicant), a minimal form is always a sum of prime implicants. Thus the first step is the calculation of the prime implicants of the function—this is discussed in § 3.5.3.

Consider now the incidence system formed with the vertices of the function $V_{0 \text{ to } v-1}$ as the objects and the prime implicants $P_{0 \text{ to } p-1}$ as the blocks. There is a cost C_i, $i \in p$, associated with each prime implicant. This may simply give the number of literal occurrences in the prime implicant, but may, in general, be arbitrary. The problem is then to find a cover with minimum cost. One approach to this problem found to be useful (see Roth and Wagner [242]) involves a branching approach (§ 6.4.1) with merged core calculation (§ 6.4.4). A program using the structure of segment B in Fig. 6.18 for the branching algorithm is given in Fig. 7.26.

The successive prime implicants on which branching is made are labeled k_h, h being the nest index. The set of prime implicant labels and vertex labels at level h of the recursive branching process are given by Q_h and W_h, respectively. The set M_h gives the elements

M.1 $Q_0 = (\text{set}) \, p; \; W_0 = (\text{set}) \, v; \; K = 10^8$

M.2 $M_0 = core(0)$ "Also changes Q_0 and W_0"

M.3   with $h = 0, 1, \ldots$, until $Q_h = \{ \; \}$:

M.4 $\vdots$

M.5 $k_h = (Q_h)_1$

M.6   for $t_h \subset \{k_h\}$:

M.7   if $t_h = \{ \; \}$:

M.8 $Q_{h+1} = Q_h \sim \{k_h\}; \; W_{h+1} = W_h; \; M_{h+1} = M_h$

M.9   otherwise:

M.10 $Q_{h+1} = Q_h \sim \{k_h\}; \; W_{h+1} = W_h \sim P_{k_h}; \; M_{h+1} = M_h \cup \{k_h\}$

M.11 $M_{h+1} \cup core(h+1) \rightarrow M_{h+1}$

M.12 $S = \sum_{i \in M_{h+1}} (C_i); \text{ if } S > K: \text{ reiterate}$

M.13 $\vdots$

M.14 $K = S; \; Min = M_h$

M.15   stop "At this point Min is a minimal cover of cost K"

C.0 $core(z)$

C.1 $Core = \{ \; \}; \; x = 1; \; y = 1$

C.2   # 1 if $x = 1$:

C.3   for $j \in W_z \ni |V_j \cap Q_z| = 1$:

C.4 $y = 1; \; i = (V_j \cap Q_z)_1$

C.5 $Q_z \sim \{i\} \rightarrow Q_z; \; W_z \sim P_i \rightarrow W_z; \; Core \cup \{i\} \rightarrow Core$

C.6 $x = 0$

C.7   if $y = 1$:

C.8   for $i \in Q_z; \; j \in Q_z \sim \{i\}$:

C.9   if $C_i \geqslant C_j$ and $(P_i \cap W_z) \subset (P_j \cap W_z)$:

C.10 $x = 1; \; Q_z \sim \{i\} \rightarrow Q_z$

C.11 $y = 0; \text{ go to } \# 1$

C.12 $core(\,) = Core;$ exit from procedure

FIG. 7.26. A program for Boolean expression minimization.

of the current partial cover; K is the cost of the best cover so far generated. The two links emanating from a node of the search tree correspond to prime implicant k_h not being included (line M.8) or being included (line M.10) in the partial cover. The core calculation is as described in § 6.4.4; the test for the first situation appears at line C.3 and the test for the second situation at lines C.8 and C.9. The tests are repeated (see line C.11) until no tests are successful and we reach line C.12.

This program is effective in finding (and proving) a minimal normal form expression. An approach useful in finding good near-minimal expressions in somewhat less time than required by this algorithm has been pursued by various authors, notably Wells [253], Edmonds [209], Ray-Chadhuri [241], and most recently Mayoh [232]. The idea is to consider the object-space consisting of all covers with a distance function based essentially on the number of blocks (prime implicants) by which two covers differ. Using a method-of-ascent scan as described in § 4.3.4, one can systematically search for better and better covers. Actually, in practice, one restricts the search to the space of irredundant covers, and a good solution is often found in the vicinity of the starting point.

EXERCISES

1. List all minimal forms for the three-variable Boolean function defined by the expression (7.9).

*2. Study improvements in the impasse detection of the program of Fig. 7.26.

GENERAL PROJECTS

1. Write a program to generate representative Boolean functions of n variables inequivalent under permutation and complementation of the variables. Slepian [247] gives the number of representatives for $n \leqslant 6$—e.g. for $n = 3$ there are twenty-two such equivalence classes.

2. Study the feasibility of constructing symmetric square zero–one matrices $B_{0 \text{ to } N-1}$ (with transpose $V_{0 \text{ to } N-1}$), where $N = 4n-1$ for given positive n, in which

$$\forall_{x \in N} (|B_x| = 2n \text{ and } |V_x| = 2n)$$

and

$$\forall_{x \in N; y \in N} (|B_x \cap B_y| = n \text{ and } |V_x \cap V_y| = n)$$

are true. Without requiring symmetry, this problem is equivalent to *Hadamard matrix* formation [255] for which constructions are known for several infinite sequences of values of n and all $n \leqslant 46$ [21]. Using techniques of this book, colleagues of the author have succeeded in constructing such a *symmetric* matrix for $n = 6$ and 7.

3. Write a program to find a set of edges of minimum cardinality whose removal eliminates all directed cycles in a given directed graph. This is the so-called Runyon problem—see Hakimi [218], for instance.

4. Let $S_1, S_2, \ldots, S_n$ be given subsets of (set) N. Write an efficient program to calculate

$$\bigcup_{i=1 \text{ to } n \ni P \subset S_i}(\{i\})$$

where P is an arbitrary subset of N. Manacher [230] has analyzed one reasonable method and shown that the number of comparisons is about

$$\frac{\pi}{2} \frac{pN}{log(|P|)},$$

where p is the probability that an S_i contains any given element of N.

5. Study the problem of enumerating the $n \times n$ zero–one matrices which have a one in every row and in every column but which contain no permutation matrices.

6. Write an efficient program to determine the possibility of "covering" an arbitrary polyomino with k given subpolyominoes. In this regard, see Fletcher [211], as well as Golomb [18].

7. Devise an algorithm for testing whether an arbitrary given polyomino is "simply connected" (i.e. contains no holes).

8. Study the problem of calculating the maximum number of vertices on the n-cube that may be distinguished in order that the minimum "distance" (along the edges of the cube) between distinguished vertices is greater than or equal to r.

9. Study the feasibility of calculating "Ramsey numbers" [37] on a computer. In particular, write a program which can be used to show that $R(3,7) > 22$. That is, write a program to search for a way to color the vertices of the complete graph with twenty-two vertices using two colors red and green in such a way that there are no 3-cliques with all red vertices or 7-cliques with all green vertices. For discussion of the theoretical calculation of Ramsey numbers, see Graver and Yackel [216].

10. Write a program to generate the polyominoes which "just fit" inside an $m \times n$ rectangle. The number of k-celled polyominoes which fit inside a 3×5 rectangle is given below:

k	7	8	9	10	11	12	13	14	15
Number	25	96	210	255	212	103	33	6	1

Parkin *et al.* [239] have computed these numbers for all m and n such that $m+n \leqslant 15$.

APPENDIX I

TABLES OF IMPORTANT NUMBERS

WE INCLUDE here short tables of some of the more important combinatorial numbers. More extensive published tables exist in the literature—see, for example, references [1] and [19]. Actually, a person doing extensive research in combinatorial computing should have ready access to many tables, and therefore should perhaps compute his own tables in a form suited to his personal taste.

TABLE I.1

n	Powers of 2 2^n	Factorials $n!$	Bell numbers (§ 3.3.5) B_n	Segner numbers (§ 5.5.4) T_n
0	1	1	1	1
1	2	1	1	1
2	4	2	2	2
3	8	6	5	5
4	16	24	15	14
5	32	120	52	42
6	64	720	203	132
7	128	5 040	877	429
8	256	40 320	4 140	1 430
9	512	362 880	21 147	4 862
10	1 024	3 628 800	115 975	16 796
11	2 048	39 916 800	678 570	58 786
12	4 096	479 001 600	4 213 597	208 012
13	8 192	6 227 020 800	27 644 437	742 900
14	16 384	87 178 291 200	190 899 322	2 674 440
15	32 768	1 307 674 368 000	1 382 958 545	9 694 845
16	65 536	20 922 789 888 000	10 480 142 147	35 357 670
17	131 072	355 687 428 096 000	82 864 869 804	129 644 790
18	262 144	6 402 373 705 728 000	682 076 806 159	477 638 700
19	524 288	121 645 100 408 832 000	5 832 742 205 057	1 767 263 190
20	1 048 576	2 432 902 008 176 640 000	51 724 158 235 372	6 564 120 420

TABLE I.2. BINOMIAL COEFFICIENTS, $\binom{n}{r} = C_{n,r}$ (EXERCISE 6 OF § 1.4)

n\r	0	1	2	3	4	5	6	7	8	9	10	11	12	13	14	15
0	1															
1	1	1														
2	1	2	1													
3	1	3	3	1												
4	1	4	6	4	1											
5	1	5	10	10	5	1										
6	1	6	15	20	15	6	1									
7	1	7	21	35	35	21	7	1								
8	1	8	28	56	70	56	28	8	1							
9	1	9	36	84	126	126	84	36	9	1						
10	1	10	45	120	210	252	210	120	45	10	1					
11	1	11	55	165	330	462	462	330	165	55	11	1				
12	1	12	66	220	495	792	924	792	495	220	66	12	1			
13	1	13	78	286	715	1287	1716	1716	1287	715	286	78	13	1		
14	1	14	91	364	1001	2002	3003	3432	3003	2002	1001	364	91	14	1	
15	1	15	105	455	1365	3003	5005	6435	6435	5005	3003	1365	455	105	15	1

(*For* TABLE I. 3 *see page* 235.)

TABLE I.4. NUMBER OF PARTITIONS OF n INTO PARTS $\leqslant m$, $R_{n,m}$ (§ 5.5.2)

n\m	0	1	2	3	4	5	6	7	8	9	10	11	12	13	14	15
0	1															
1	0	1														
2	0	1	2													
3	0	1	2	3												
4	0	1	3	4	5											
5	0	1	3	5	6	7										
6	0	1	4	7	9	10	11									
7	0	1	4	8	11	13	14	15								
8	0	1	5	10	15	18	20	21	22							
9	0	1	5	12	18	23	26	28	29	30						
10	0	1	6	14	23	30	35	38	40	41	42					
11	0	1	6	16	27	37	44	49	52	54	55	56				
12	0	1	7	19	34	47	58	65	70	73	75	76	77			
13	0	1	7	21	39	57	71	82	89	94	97	99	100	101		
14	0	1	8	24	47	70	90	105	116	123	128	131	133	134	135	
15	0	1	8	27	54	84	110	131	146	157	164	169	172	174	175	176

TABLE I.3. STIRLING NUMBERS OF SECOND KIND, $S_{n,k}$ (EXERCISE 1 OF § 2.3)

k \ n	1	2	3	4	5	6	7	8	9	10	11	12	13	14
1	1													
2	1	1												
3	1	3	1											
4	1	7	6	1										
5	1	15	25	10	1									
6	1	31	90	65	15	1								
7	1	63	301	350	140	21	1							
8	1	127	966	1 701	1 050	266	28	1						
9	1	255	3 025	7 770	6 951	2 646	462	36	1					
10	1	511	9 330	34 105	42 525	22 827	5 880	750	45	1				
11	1	1023	28 501	145 750	246 730	179 487	63 987	11 880	1 155	55	1			
12	1	2047	86 526	611 501	1 379 400	1 323 652	627 396	159 027	22 275	1 705	66	1		
13	1	4095	261 625	2 532 530	7 508 501	9 321 312	5 715 424	1 899 612	359 502	39 325	2 431	78	1	
14	1	8191	788 970	10 391 745	40 075 035	63 436 373	49 329 280	20 912 320	5 135 130	752 752	66 066	3367	91	1

TABLE I.5. PRIME NUMBERS < 1000 (BY BIT-PATTERN)

	00					10					20					30					40					50					60					70					80					90				
	1	3	5	7	9	1	3	5	7	9	1	3	5	7	9	1	3	5	7	9	1	3	5	7	9	1	3	5	7	9	1	3	5	7	9	1	3	5	7	9	1	3	5	7	9	1	3	5	7	9
0	(1)	1	1	1	0	1	1	0	1	1	0	1	0	0	1	1	0	0	1	0	1	1	0	1	0	0	1	0	0	1	1	0	0	1	0	1	1	0	0	1	0	1	0	0	1	0	0	0	1	0
100	1	1	0	1	1	0	1	0	0	0	0	0	0	1	0	1	0	0	1	1	0	0	0	0	1	1	0	0	1	0	0	1	0	1	0	0	1	0	0	1	1	0	0	0	0	1	1	0	1	1
200	0	0	0	0	0	1	0	0	0	0	0	1	0	1	1	0	1	0	0	1	1	0	0	0	0	1	0	0	1	0	0	1	0	0	1	1	0	0	1	0	1	1	0	0	0	0	1	0	0	0
300	0	0	0	1	0	1	1	0	1	0	0	0	0	0	0	1	0	0	1	0	0	0	0	1	1	0	1	0	0	1	0	0	0	1	0	0	1	0	0	1	0	1	0	0	1	0	0	0	1	0
400	1	0	0	0	1	0	0	0	0	1	1	0	0	0	0	1	1	0	0	1	0	1	0	0	1	0	0	0	1	0	1	1	0	1	0	0	0	0	0	1	0	0	0	1	0	1	0	0	0	1
500	0	1	0	0	1	0	0	0	0	0	1	1	0	0	0	0	0	0	0	0	1	0	0	1	0	0	0	0	1	0	0	1	0	0	1	1	0	0	1	0	0	0	0	1	0	0	1	0	0	1
600	1	0	0	1	0	0	1	0	1	1	0	0	0	0	0	1	0	0	0	0	1	1	0	1	0	0	1	0	0	1	1	0	0	0	0	0	1	0	1	0	0	1	0	0	0	1	0	0	0	0
700	1	0	0	0	1	0	0	0	0	1	0	0	0	1	0	0	1	0	0	1	0	1	0	0	0	1	0	0	1	0	1	0	0	0	1	0	1	0	0	0	0	0	0	1	0	0	0	0	1	0
800	0	0	0	0	1	1	0	0	0	0	1	1	0	1	1	0	0	0	0	1	0	0	0	0	0	0	1	0	1	1	0	1	0	0	0	0	0	0	1	0	1	1	0	1	0	0	0	0	0	0
900	0	0	0	1	0	1	0	0	0	1	0	0	0	0	1	0	0	0	1	0	1	0	0	1	0	0	1	0	0	0	0	0	0	1	0	1	0	0	1	0	0	1	0	0	0	1	0	0	1	0

APPENDIX II

TABLES OF INTERESTING NUMBERS

WE INCLUDE here tables of certain combinatorial numbers computed by the author. We believe this is the first time that Tables II.6 and II.7 have appeared in print.

TABLE II.1. EXCESS NUMBER OF MULTIPLES ($< 2^n$) OF p WITH EVEN 1-BIT PARITY (§ 3.1.5)

n	$p = 3$	$p = 7$	$p = 11$	$p = 17$
1	1	1	1	1
2	2	1	1	1
3	3	0	1	1
4	6	−1	0	1
5	9	−3	−1	2
6	18	−6	−2	4
7	27	−7	−2	8
8	54	−7	0	16
9	81	0	5	21
10	162	7	10	29
11	243	21	11	37
12	486	42	11	37
13	729	49	11	74
14	1 458	49	0	140
15	2 187	0	−11	264
16	4 374	− 49	−22	528
17	6 561	−147	−22	697
18	13 122	−294	0	969
19	19 683	−343	55	1 241
20	39 366	−343	110	1 241
21	59 049	0	121	2 482
22	118 098	343	121	4 692
23	177 147	1 029	121	8 840
24	354 294	2 058	0	17 680
25	531 441	2 401	−121	23 341
26	1 062 882	2 401	−242	32 453
27	1 594 323	0	−242	41 565
28	3 188 646	−2 401	0	41 565
29	4 782 969	−7 203	605	83 130
30	9 565 938	−14 406	1210	157 148
31	14 348 907	−16 807	1331	296 072
32	28 697 814	−16 807	1331	592 144
33	43 046 721	0	1331	781 745
34	86 093 442	16 807	0	1 086 929
35	129 140 163	50 421	−1331	1 392 113

TABLE II.2. NUMBER OF SOLUTIONS
TO QUEENS' PROBLEM
(EXERCISE 1 OF § 5.2)
(EXERCISE 4 OF § 6.5)

n	q_n	Iq_n
1	1	1
2	0	0
3	0	0
4	2	1
5	10	2
6	4	1
7	40	6
8	92	12
9	352	46
10	724	92
11	2 680	341
12	14 200	1787
13	73 712	9233

For an $n \times n$ chessboard, q_n is the total number of ways to place n nonattacking queens on the board. Iq_n is the number of ways inequivalent under the group of the square.

TABLE II.3. FOLDING NUMBERS
(SEE EXERCISE 13 OF § 5.2)

n	f_n	g_n
1	1	
2	2	
3	6	1
4	16	2
5	50	5
6	144	12
7	462	33
8	1 392	87
9	4 536	252
10	14 060	703
11	46 310	2 105
12	146 376	6 099
13	485 914	18 689

TABLE II.4. ANSWERS TO
A DISTRIBUTION PROBLEM
FOR $n = 5$
(EXERCISE 14 OF § 5.5)

Partition p	$N(p)$
1111111111	945
211111111	525
22111111	300
31111111	210
2221111	177
3211111	125
222211	109
322111	77
4111111	75
22222	73
331111	56
32221	50
421111	47
33211	36
42211	31
511111	26
3322	25
4222	23
43111	23
3331	18
52111	17
4321	16
5221	12
4411	11
61111	10
442	9
433	9
5311	9
6211	7
532	7
622	6
541	5
631	4
7111	4
64	3
721	3
55	3
811	2
73	2
82	2
91	1
10	1

$N(p)$ is the number of distinct ways in which ten objects of specification p can be placed in five indistinguishable boxes, each box containing exactly two objects.

TABLE II.5. NUMBER OF REDUCED
LATIN SQUARES (§ 7.1.3)

n	l_n	Il_n
3	1	1
4	4	2
5	56	2
6	9 408	12
7	16 942 080	147
8	535 281 401 856	

The quantity l_n is the number of reduced Latin squares of order n, while Il_n is the number of squares inequivalent under label, row, and/or column permutations and permutation of labels with rows, labels with columns, etc. (The group has order $6n!^3$.)

TABLE II.6. NUMBER OF ZERO–ONE MATRICES WITH VANISHING
PERMANENT (EXERCISE 8 OF § 6.4)

2×2		3×3		4×4		5×5	
k	No.	k	No.	k	No.	k	No.
2	4	3	6	4	8	5	10
3	4	4	45	5	96	6	200
4	1	5	90	6	576	7	1 900
		6	78	7	2128	8	11 500
		7	36	8	4860	9	50 025
		8	9	9	6976	10	166 720
		9	1	10	6496	11	439 600
				11	4080	12	923 700
				12	1796	13	1 534 800
				13	560	14	1 994 200
				14	120	15	2 010 920
				15	16	16	1 571 525
				16	1	17	956 775
						18	458 500
						19	174 700
						20	53 010
						21	12 650
						22	2 300
						23	300
						24	25
						25	1

The column labeled k gives the number of zeros in the $n \times n$ matrix. The No. column then gives the number of $n \times n$ zero–one matrices containing k zeros (hence $n^2 - k$ ones) whose permanent is zero.

TABLE II.7. SPECTRA OF DETERMINANT MAGNITUDES FOR ZERO–ONE MATRICES
(EXERCISE 3 OF § 4.4 AND EXERCISE 7 OF § 7.1)

$n = 3$

M	N
0	27
1	28
2	1

$n = 4$

M	N
0	880
1	835
2	100
3	5

$n = 5$

M	N
0	97 090
1	80 856
2	20 232
3	2 412
4	726
5	60

$n = 6$

M	N
0	34 923 518
1	25 666 809
2	10 746 288
3	2 135 343
4	1 163 064
5	176 701
6	129 360
7	17 885
8	13 930
9	1 470

$n = 7$

M	N
0	40 885 781 314
1	26 883 246 720
2	16 511 989 560
3	4 650 079 360
4	3 511 706 880
5	744 944 448
6	833 612 648
7	161 359 296
8	208 846 176
9	57 084 608
10	42 833 560
11	9 880 640
12	17 749 760
13	2 437 120
14	2 432 640
15	806 400
16	759 360
17	80 640
18	135 240
19	0
20	26 880
21	0
22	0
23	0
24	1 920
25	0
26	0
27	0
28	0
29	0
30	0
31	0
32	30

The columns labeled M give the "possible" values for the magnitude of the determinant of an $n \times n$ matrix. The columns labeled N give the number of matrices with distinct ordered rows whose determinant has the corresponding magnitude.

APPENDIX III

COMPENDIUM OF FOUR-COLOR REDUCIBLE CONFIGURATIONS

THE reducible configurations (§ 7.3) known to this author are given in Fig. III.1. They include all those given in chapter 12 of Ore [235] and some new ones calculated by the author.

The numbers attached to the vertices are the potential valences. Each of the configurations given here remains reducible when one or more of the potential valences are replaced by smaller whole numbers.

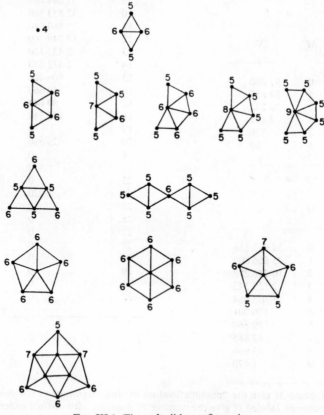

FIG. III.1. The reducible configurations.

BIBLIOGRAPHY

GENERAL

[1] ABRAMOWITZ, M. and STEGUN, I. A. (editors), *Handbook of Mathematical Functions with Formulas, Graphs, and Mathematical Tables*, National Bureau of Standards, Washington, D.C., 1964.

[2] BALL, W. W. R., *Mathematical Recreation and Essays*, revised edition, Macmillan, New York, 1962.

[3] BECKENBACH, E. F. (editor), *Applied Combinatorial Mathematics*, John Wiley, New York, 1964.

[4] BELLMAN, R. E. (editor), *Proceedings of Symposia in Applied Mathematics 14: Mathematical Problems in the Biological Sciences*, American Mathematical Society, Providence, R.I., 1962.

[5] BELLMAN, R. E., and DREYFUS, S. E., *Applied Dynamic Programming*, Princeton U.P., Princeton, N.J., 1962.

[6] BELLMAN, R. E. and HALL, M., JR., (editors), *Proceedings of Symposia in Applied Mathematics 10: Combinatorial Analysis*, American Mathematical Society, Providence, R.I., 1960.

[7] BUSACKER, R. G. and SAATY, T. L., *Finite Graphs and Networks*, McGraw-Hill, New York, 1965.

[8] CANNON, J., Computers in group theory: a survey, *Comm. ACM* **12**, 3 (1969).

[9] CARMICHAEL, R. D., *Introduction to the Theory of Groups of Finite Order*, Dover, New York, 1956.

[10] CHURCHHOUSE, R. F. and HERZ, J. C. (editors), *Computers in Mathematical Research*, North-Holland, Amsterdam, 1968.

[11] *Communications of ACM, Algorithms*, Association for Computing Machinery, Baltimore, Md., 1958–.

[12] COXETER, H. S. M. and MOSER, W. O. J., *Ergebnisse der Mathematik und ihrer Grenzgebiete 14: Generators and Relations for Discrete Groups*, Springer-Verlag, New York, 1965.

[13] CURTISS, J. H. (editor), *Proceedings of Symposia in Applied Mathematics 6: Numerical Analysis*, McGraw-Hill for American Mathematical Society, Providence, R.I., 1956.

[14] DÉNES, J., A bibliography on non-numerical applications of digital computers, *Comput. Rev.* **9**, 481 (1968).

[15] FREIBERGER, W. F. and PRAGER, W. (editors), *Applications of Digital Computers*, Ginn, Boston, 1963.

[16] FULKERSON, D. R., Flow networks and combinatorial operations research, *Amer. Math. Monthly* **73**, 115 (1966).

[17] GOLOMB, S. W., Mathematical theory of discrete classification, in *Information Theory, Fourth London Symposium*, C. Cherry (editor), Butterworths, Washington, D.C., 1961.

[18] GOLOMB, S. W., *Polyominoes*, Charles Scribner's, New York, 1965.

[19] GUPTA, H., GWYTHER, C. E. and MILLER, J. C. P., *Royal Society Mathematical Tables 4: Tables of Partitions*, Cambridge U.P., London, 1958.

[20] HALL, M., JR., A survey of combinatorial analysis, chapter 3 in *Some Aspects of Analysis and Probability IV*, John Wiley, New York, 1958.

[21] HALL, M., JR., *Combinatorial Theory*, Blaisdell, Waltham, Mass., 1967.

[22] HALL, M., JR. and KNUTH, D. E., Combinatorial analysis and computers, *Amer. Math. Monthly* **72**, Part 2, 21 (1965).

[22A] HARARY, F. (editor), *Proof Techniques in Graph Theory*, Academic, New York, 1969.

[23] *IBM Scientific Computing Symposium on Combinatorial Problems, Proceedings, March 16–18, 1964*, IBM, White Plains, 1966.

[24] KAUFMANN, A., *Graphs, Dynamic Programming and Finite Games*, Academic, New York, 1967.

[25] KEMENY, J. G., SNELL, J. L. and THOMPSON, G. L., *Introduction to Finite Mathematics*, Prentice-Hall, Englewood Cliffs, N.J., 1957.

[26] KLERER, M. and KORN, G. A. (editors), *Digital Computer User's Handbook*, McGraw-Hill, New York, 1967.

[27] KNUTH, D. E., *The Art of Computer Programming*, Volume 1: *Fundamental Algorithms*, Addison-Wesley, Reading, Mass., 1968.

[28] LIU, C. L., *Introduction to Combinatorial Mathematics*, McGraw-Hill, New York, 1968.

[29] MACMAHON, P. A., *Combinatory Analysis*, Chelsea, New York, 1960.

[30] *Mathematical Algorithms*, MIT, Cambridge, Mass., 1966–.

[31] METROPOLIS, N., TAUB, A. H., TODD, J. and TOMPKINS, C. B. (editors), *Proceedings of Symposia in Applied Mathematics 15: Experimental Arithmetic, High-Speed Computing and Mathematics*, American Mathematical Society, Providence, R.I., 1963.

[32] NETTO, E., *Lehrbuch der Combinatorik*, second edition, Chelsea, New York, 1927.

[33] NIVEN, J., *Mathematics of Choice—How to Count Without Counting*, Random House, New York, 1965.

[34] ORE, O., *Number Theory and its History*, McGraw-Hill, New York, 1948.

[35] PRATHER, R. E., *Introduction to Switching Theory: A Mathematical Approach*, Allyn & Bacon, Boston, Mass., 1967.

[36] RIORDAN, J., *An Introduction to Combinatorial Analysis*, John Wiley, New York, 1958.

[37] RYSER, H. J., *Combinatorial Mathematics*, Mathematical Association of America, Buffalo, N.Y., 1963.

[38] SAMMET, J. E., An annotated descriptor based bibliography on the use of computers for non-numeric mathematics, *Comput. Rev.* **7**, B1 (1966).

[39] RALSTON, A. and WILF, H. S. (editors), *Mathematical Methods for Digital Computers*, Volume 2, John Wiley, New York, 1967.

[40] SCHWARTZ, J. T. (editor), *Proceedings of Symposia in Applied Mathematics 19: Mathematical Aspects of Computer Science*, American Mathematical Society, Providence, R.I., 1967.

[41] SESHU, S. and REED, M. B., *Linear Graphs and Electrical Networks*, Addison-Wesley, Reading, Mass., 1961.

[42] TODD, J. (editor), *A Survey of Numerical Analysis* (Chapters 15 and 16), McGraw-Hill, New York, 1962.

[43] TUCKER, A. W. *et al.*, Combinatorial problems, *IBM J. Res. Develop.* **4**, supplementary issue (1960).

CHAPTER 1

[44] GALLER, B. A., *The Language of Computers*, McGraw-Hill, New York, 1962.

[45] HIGMAN, B., *A Comparative Study of Programming Languages*, American Elsevier, New York, 1967.

[46] IVERSON, K. E., *A Programming Language*, John Wiley, New York, 1962.

[47] LECHT, C. P., *The Programmer's PL/1, a Complete Reference*, McGraw-Hill, New York, 1968.

[48] ORGANICK, E. I., *A Fortran IV Primer*, Addison-Wesley, Reading, Mass., 1966.

[49] ROSEN, S. (editor), *Programming Systems and Languages*, McGraw-Hill, New York, 1967.

[50] RUTISHAUSER, H., The use of recursive procedures in Algol 60, page 43 in *Advances in Computers 3*, Academic, New York, 1963.

[51] SAMMET, J. E., *Programming Languages: History and Fundamentals*, Prentice-Hall, Englewood Cliffs, N.J., 1968.

[52] STEIN, J., Computational problems associated with Racah algebra, *J. Computational Phys.* **1**, 397 (1967).

[53] WELLS, M. B., Recent improvements in Madcap, *Comm. ACM* **6**, 674 (1963).

[54] WIRTH, N. and HOARE, C. A. R., A contribution to the development of Algol, *Comm. ACM* **9**, 413 (1966).

[55] WOOLRIDGE, R. and RACTLIFFE, J. F., *An Introduction to Algol Programming*, English Universities Press, London, 1963.

CHAPTER 2

[56] DAVIS, R., Programming language processors, page 117 in *Advances in Computers 7*, Academic, New York, 1966.

[57] DIJKSTRA, E. W., Recursive programming, *Numer. Math.* **2,** 312 (1960).
[58] HASTINGS, C., HAYWARD, J. T. and WONG, J. P., *Approximations for Digital Computers*, Princeton U.P., Princeton, N.J., 1955.
[59] HULL, T. E. and DOBELL, A. R., Random number generators, *SIAM Rev.* **4,** 230 (1962).
[60] HUTCHINSON, D., A new uniform pseudo-random number generator, *Comm. ACM* **9,** 432 (1966).
[61] JANSSON, B., *Random Number Generators*, Almquist & Wiksell, Stockholm, 1966.
[62] LAZARUS, R., WELLS, M. B. and WOOTEN, J., JR., The Maniac II system, in *Proceedings of a Symposium on Interactive Systems for Experimental Applied Mathematics*, M. Klerer and J. Reinfelds (editors), Academic, New York, 1968.
[63] LEHMER, D. H., Mathematical methods in large-scale computing units, in *Annals Comp. Lab. Harvard U.* **26,** 141 (1951).
[64] LEHMER, D. H., Teaching combinatorial tricks to a computer, page 179 in reference [6].
[65] LYUSTERNIK, L. A., CHERVONENKIS, O. A. and YANPOL'SKII, A. R., *Handbook for Computing Elementary Functions*, Pergamon, New York, 1965.
[66] RANDELL, B. and KUEHNER, C. J., Dynamic storage allocation systems, *Comm. ACM* **11,** 297 (1968).
[67] RIESEL, H., In which order are different conditions to be examined?, *BIT* **3,** 255 (1963).
[68] SLAGLE, J. R., An efficient algorithm for finding certain minimum-cost procedures for making binary decisions, *J. Assoc. Comput. Mach.* **11,** 253 (1964).
[69] STEMMLER, R. M., The ideal Waring theorem for exponents 401–200000, *Math. Comp.* **18,** 144 (1964).
[70] TIENARI, M. and SUOKONAUTEO, V., A set of procedures making real arithmetic of unlimited accuracy possible within Algol 60, *BIT* **6,** 332 (1966).
[71] WELLS, M. B., Aspects of language design for combinatorial computing, *IEEE Trans. Electronic Computers* **EC-13,** 431 (1964).

CHAPTER 3

[72] COMÉT, S., Factorization of factorials, *BIT* **1,** 167 (1961).
[73] COMÉT, S., Notations for partitions, *Math. Tables Aids Comput.* **9,** 143 (1955).
[74] ELSPAS, B., The theory of autonomous linear sequential network, *IRE Trans. on Circuit Theory* **6,** 45 (1959).
[75] ERDŐS, P., GOODMAN, A. W. and PÓSA, L., The representation of a graph by set intersections, *Canad. J. Math.* **18,** 106 (1966).
[76] FORSYTHE, G. E., SWAC computes 126 distinct semigroups of order 4, *Proc. Amer. Math. Soc.* **6,** 443 (1955).
[77] FLORES, I., Note on a machine algorithm for conversion from reflected binary to natural binary, *Comput. J.* **7,** 121 (1964).
[78] FOX, L. (editor), *Advances in Programming and Non-numerical Computation*, Pergamon, New York, 1966.
[79] GARNER, H. L., Number systems and arithmetic, page 131 in *Advances in Computers* **6,** Academic, New York, 1965.
[80] GIOIA, A. A., SUBBARAO, M. V. and SUGUNAMMA, M., The Scholz–Brauer problem in addition chains, *Duke Math. J.* **29,** 481 (1962).
[81] GROSSMAN, J., *Groups and Their Graphs*, Random House, New York, 1966.
[82] HUTCHINSON, G., Partitioning algorithms for finite sets, *Comm. ACM* **6,** 613 (1963).
[83] *Journal of Research, National Bureau of Standards* **69 B,** Numbers 1 and 2 (1965).
[84] KNUTH, D. E., *The Art of Computer Programming*, Volume 2: *Semi-numerical Algorithms*, Addison-Wesley, Reading, Mass., 1969.
[85] LEECH, J., Coset enumeration on digital computers, *Proc. Cambridge Philos. Soc.* **59,** 257 (1963).
[86] LEHMER, D. H., Machine tools of combinatorics, Chapter 1 in reference [3].
[87] LITTLEWOOD, D. E., *Theory of Group Characters and Matrix Representations of Groups*, Clarendon Press, Oxford, 1950.
[88] MANN, H. B., On modular computation, *Math. Comp.* **15,** 190 (1961).
[89] McCLUSKEY, E. J., JR., Minimization of Boolean functions, *Bell System Tech. J.* **35,** 1417 (1956).
[90] MILETO, F. and PUTZOLU, G., Average values of quantities appearing in Boolean function minimization, *IEEE Trans. Electronic Computers* **EC-13,** 87 (1964).

[91] MILLER, J. C. P. and BROWN, D. J. S., An algorithm for evaluation of remote terms in a linear recurrence sequence, *Comput. J.* **9**, 188 (1966).

[92] PETERSON, W. W., *Error-Correcting Codes*, Technology Press and John Wiley, Cambridge, Mass. and New York, 1961.

[93] PFALTZ, J. L. and ROSENFELD, A., Computer representation of planar regions by their skeletons, *Comm. ACM* **10**, 119 (1967).

[94] QUINE, W. V., The problem of simplifying truth functions, *Amer. Math. Monthly* **59**, 521 (1952).

[95] ROTENBERG, M., BIVINS, R., METROPOLIS, N., WOOTEN, J. K., JR., *The 3-j and 6-j Symbols*, Technology Press, MIT, Cambridge, Mass., 1959.

[96] SWIFT, J. D., Construction of Galois fields of characteristic two and irreducible polynomials, *Math. Comp.* **14**, 99 (1960).

[97] SZABÓ, N. S. and TANAKA, R. I., *Residue Arithmetic and its Application to Computer Technology*, McGraw-Hill, New York, 1967.

[98] TROTTER, H. F., A machine program for coset enumeration, *Canad. Math. Bull.* **7**, 357 (1964).

[99] ULAM, S. M., On some mathematical problems connected with patterns of growth of figures, page 215 in reference [4].

[100] WELLS, M. B., Bit parity of multiples of primes, *Abstracts of Short Communications*, International Congress of Mathematicians, Stockholm, 1962.

[101] WELLS, M. B., Simplification of normal form expressions for Boolean functions of many variables, Doctoral Thesis, Univ. of Calif., 1961.

[102] WUNDERLICH, M., Certain properties of pyramidal and figurate numbers, *Math. Comp.* **16**, 482 (1962).

CHAPTER 4

[103] BELLMAN, R.E., Combinatorial processes and dynamic programming, page 217 in reference [6].

[104] BERLEKAMP, E. R., Program for double-dummy bridge problems—a new strategy for mechanical game playing, *J. Assoc. Comput. Mach.* **10**, 357 (1963).

[105] COLLINS, N. L. and MICHIE, D. (editors), *Machine Intelligence* 1, American Elsevier, New York, 1967.

[106] DOUGLAS, R. J., Some results on the maximum length of circuits of spread k in the d-cube, *J. of Combinatorial Theory* **6**, 323 (1969).

[107] FLOYD, R. W., Nondeterministic algorithms, *J. Assoc. Comput. Mach.* **14**, 636 (1967).

[108] GLEASON, A. M., A search problem in the n-cube, page 175 in reference [6].

[109] GLOVER, F., Some truncated-enumeration methods for solving integer linear programs, preprint of paper presented at 29th National Meeting of Operations Research Society of America, 1966.

[110] GOLOMB, S. W., Efficient coding for the desoxyribonucleic channel, page 87 in reference [4].

[111] GOLOMB, S. W. and BAUMERT, L. D., Backtrack programming, *J. Assoc. Comput. Mach.* **12**, 516 (1965).

[112] GOMORY, R. E., The traveling salesman problem, page 93 in reference [23].

[113] HELD, M. and KARP, R. M., A dynamic programming approach to sequencing problems, *SIAM J. Appl. Math.* **10**, 196 (1962).

[114] HELD, M. and KARP, R. M., The construction of discrete dynamic programming algorithms, *IBM Systems J.* **4**, 136 (1965).

[115] JOHANSEN, P., Non-deterministic programming, *BIT* **7**, 289 (1967).

[116] JURKAT, W. B. and RYSER, H. J., Matrix factorizations of determinants and permanents, *J. Algebra* **3**, 1 (1966).

[117] KLEE, V., A method for constructing circuit codes, *J. Assoc. Comput. Mach.* **14**, 520 (1967).

[118] KUHN, H. W., The Hungarian method for the assignment problem, *Naval Res. Logistics Quarterly* **2**, 83 (1955).

[119] LAWLER, E. L. and BELL, M. D., A method for solving discrete optimization problems, *Operations Res.* **14**, 1098 (1966).

[120] LAWLER, E. L. and WOOD, D. E., Branch-and-bound methods: a survey, *Operations Res.* **14**, 699 (1966).

[121] LEHMER, D. H., Teaching combinatorial tricks to a computer, page 179 in reference [6].

[122] LEHMER, D. H., Some high-speed logic, page 141 in reference [31].

[123] LIN, S., Computer solutions of the traveling salesman problem, *Bell System Tech. J.* **44**, 2245 (1965).

[124] LITTLE, J. D. C., MURTY, K. G., SWEENEY, D. W. and KAREL, C., An algorithm for the traveling salesman problem, *Operations Res.* **11**, 972 (1963).

[125] LYNCH, W. C., Recursive solution of a class of combinatorial problems: an example., *Comm. ACM* **8**, 617 (1965).

[126] MUNKRES, J., Algorithms for the assignment and transportation problems, *SIAM J. Appl. Math.* **5**, 32 (1957).

[127] POHL, I., A method for finding Hamiltonian paths and knight's tours, *Comm. ACM* **10**, 446 (1967). See also DUBY, J. J., Improved Hamiltonian paths (letter to the editor), *Comm. ACM* **11**, 1 (1967).

[128] READ, R. C., Teaching graph theory to a computer, page 161 in *Recent Progress in Combinatorics*, TUTTE, W. T. (editor), Academic, New York, 1969.

[129] REITER, S. and SHERMAN, G., Discrete optimizing, *SIAM J. Appl. Math.* **13**, 864 (1965).

[130] ROSSMAN, M. J. and TWERY, R. J., Combinatorial programming, preprint of paper presented at 6th Annual Meeting of Operations Research Society of America, 1958.

[131] SAMUEL, A. L., Programming computers to play games, page 165 in *Advances in Computers* **1**, Academic, New York, 1960.

[132] SAMUEL, A. L., Some studies in machine learning using the game of checkers: II—recent progress, *IBM J. of Res. and Develop.* **11**, 601 (1967).

[133] WALKER, R. J., An enumerative technique for a class of combinatorial problems, page 91 in reference [6].

[134] YAMABE, H. and POPE, D., A computational approach to the four-color problem, *Math. Comp.* **15**, 250 (1961).

CHAPTER 5

[135] BRATLEY, P. and MCKAY, J. K. S., Multi-dimensional partition generator—Algorithm 313, *Comm. ACM* **10**, 666 (1967).

[136] BROWN, W. G., Historical note on a recurrent combinatorial problem, *Amer. Math. Monthly* **72**, 973 (1965).

[137] DE BALBINE, G., Note on random permutations, *Math. Comp.* **21**, 710 (1967).

[138] DE BRUIJN, N. G. and MORSELT, B. J. M., A note on plane trees, *J. Combinatorial Theory* **2**, 27 (1967).

[139] FISCHER, L. L. and KRAUSE, K. CHR., *Lehrbuch der Combinationslehre und der Arithmetik*, Dresden, 1812.

[140] HARARY, F., Permutations with restricted positions, *Math. Comp.* **16**, 222 (1962).

[141] HEAP, B. P., Permutations by interchanges, *Comput. J.* **6**, 293 (1963).

[142] JOHNSON, S. M., Generation of permutations by adjacent transposition, *Math. Comp.* **17**, 282 (1963).

[143] LEHMER, D. H., Combinatorial types in number-theory calculations, page 23 in reference [23].

[144] LEHMER, D. H., Permutation by adjacent interchanges, *Amer. Math. Monthly* **72**, Part 2, 21 (1965).

[145] LUNNON, W. F., A map-folding problem, *Math. Comp.* **22**, 193 (1968).

[146] OSTROWSKI, R. T. and VAN DUREN, K. D., On a theorem of Mann on Latin squares, *Math. Comp.* **15**, 293 (1961).

[147] PAPWORTH, D. G., Computers and change-ringing, *Comput. J.* **3**, 47 (1960).

[148] SHEN, MOK-KONG, On the generation of permutations and combinations, *BIT* **2**, 228 (1962).

[149] TOMPKINS, C. B., Machine attacks on problems whose variables are permutations, page 195 of reference [13].

[150] TROTTER, H. F., Algorithm 115—Perm, *Comm. ACM* **5**, 434 (1962).

[151] WAITE, W. M., An efficient procedure for the generation of closed subsets, *Comm. ACM* **10**, 169 (1967).

[152] WELLS, M. B., Generation of permutations by transposition, *Math. Comp.* **15**, 192 (1961).

CHAPTER 6

[153] BELL, D. A., The principles of sorting, *Comput. J.* **1**, 71 (1958).

[154] BERMAN, M. F., A method for transposing a matrix, *J. Assoc. Comput. Mach.* **5**, 383 (1958).

[155] BIRKHOFF, G., *Lattice Theory*, American Mathematical Society, Providence, R.I., 1967.

[156] BRILLHART, J. and SELFRIDGE, J. L., Some factorizations of $2^n \pm 1$ and related results, *Math. Comp.* **21**, 87 (1967).

[157] CANTOR, D. G., ESTRIN, G., FRAENKEL, A. S. and TURN, R., A very high-speed digital number sieve, *Math. Comp.* **16**, 141 (1962).

[158] FLORES, I., Analysis of internal sorting, *J. Assoc. Comput. Mach.* **8**, 41 (1961).

[159] FLORES, I., Computer time for address calculation sorting, *J. Assoc. Comput. Mach.* **7**, 389 (1960).

[160] FRANK, R. M. and LAZARUS, R. B., A high speed sorting procedure, *Comm. ACM* **3**, 20 (1960).

[161] GARDINER, V., LAZARUS, R., METROPOLIS, N. and ULAM, S., On certain sequences of integers defined by sieves, *Math. Mag.* **29**, 117 (1956).

[162] GILBERT, E. N., Gray codes and paths on the n-cube, *Bell System Tech. J.* **37**, 815 (1958).

[163] GOLDSTINE, H. H. and VON NEUMANN, J., *Planning and Coding of Problems for an Electronic Computing Instrument*, Part II, Vol. 3, Chapter 5 in *John von Neumann, Collected Works*, Volume 5, A. H. Taub (editor), Macmillan, New York, 1963.

[164] GOTLIEB, C. C., Sorting on computers, *Comm. ACM* **6**, 195 (1963). The entire May 1963 issue of *Comm. ACM* is devoted to papers presented at ACM Sort Symposium, November, 1962.

[165] GOTLIEB, C. C., Sorting on computers, page 68 in reference [15].

[166] HALL, M., JR., An algorithm for distinct representatives, *Amer. Math. Monthly* **63**, 716 (1956).

[167] HALL, M., JR. and SWIFT, J. D., Determination of Steiner triple systems of order 15, *Math. Tables Aids Comput.* **9**, 146 (1955).

[168] HIBBARD, T. H., Some combinatorial properties of certain trees with applications to searching and sorting, *J. Assoc. Comput. Mach.* **9**, 13 (1962).

[169] HILDEBRANDT, P. and ISBITZ, H., Radix exchange—an internal sorting method for digital computers, *J. Assoc. Comput. Mach.* **6**, 156 (1959).

[170] ISAAC, E. J. and SINGLETON, R. C., Sorting by address calculation, *J. Assoc. Comput. Mach.* **3**, 169 (1956).

[171] KAMPS, H. J. L. and VAN LINT, J. H., The football pool problem for 5 matches, *J. Combinatorial Theory* **3**, 309 (1967).

[172] LAWLER, E. L., Covering problems: Duality relations and a new method of solution, *SIAM J. Appl. Math.* **14**, 1115 (1966).

[173] LEHMER, D. H., An announcement concerning the delay line SIEVE DLS-127, *Math. Comp.* **20**, 645 (1966).

[174] LEHMER, D. H., The sieve problem for all-purpose computers, *Math. Tables Aids Comput.* **7**, 6 (1953).

[175] LEHMER, D. H., LEHMER, E., MILLS, W. H. and SELFRIDGE, J. L., Machine proof of a theorem on cubic residues, *Math. Comp.* **16**, 407 (1962).

[176] MARIMONT, R. B., A new method for checking the consistency of precedence matrices, *J. Assoc. Comput. Mach.* **6**, 164 (1959).

[177] MCCLUSKEY, E. J., JR., Minimization of Boolean functions, *Bell System Tech. J.* **35**, 1417 (1956).

[178] PALL, G. and SEIDEN, E., A problem in Abelian groups, with application to the transposition of a matrix on an electronic computer, *Math. Tables Aids Comput.* **14**, 189 (1960).

[179] PLEMMONS, R. J., On computing non-equivalent finite algebraic systems, *Math. Algorithms* **2**, 80 (1967).

[180] POLLACK, M., Solutions of the k-th best route through a network—a review, *J. Math. Anal. Applications* **3**, 547 (1961).

[181] POLLACK, M. and WIEBENSON, W., Solutions of the shortest route problem—a review, *Operations Res.* **8**, 224 (1960).

[182] ROTH, J. P., Algebraic topological methods in synthesis, in *Proceedings of an International Symposium on the Theory of Switching*, Harvard U., Cambridge, Mass., 1957.

[183] SALZER, H. E. and LEVINE, N., Tables of integers not exceeding 1000000 that are not expressible as the sum of 4 tetrahedral numbers, *Math. Tables Aids Comput.* **12**, 141 (1958).

[184] SHELL, D. L., A high speed sorting procedure, *Comm. ACM* **2**, 30 (1959).

[185] SHEN, MOK-KONG, On checking the Goldbach conjecture, *BIT* **4**, 243 (1964).

[186] SPENCER, J., Maximal consistent families of triples, *J. Combinatorial Theory* **5**, 1 (1968).

[187] STEIN, M. L. and STEIN, P. R., Experimental results on additive 2-bases, *Math. Comp.* **19**, 427 (1965).

[188] STEIN, M. L. and STEIN, P. R., New experimental results on the Goldbach conjecture, *Math. Mag.* **38,** 72 (1965).

[189] SWIFT, J. D., Isomorph rejection in exhaustive search techniques, page 195 in reference [6].

[190] WARSHALL, S., A theorem on Boolean matrices, *J. Assoc. Comput. Mach.* **9,** 11 (1962).

[191] WATSON, E. J., Primitive polynomials (mod 2), *Math. Comp.* **16,** 368 (1962).

[192] WELLS, M., Enlargement of a group—Algorithm 136, *Comm. ACM* **5,** 555 (1962).

[193] WHITE, A. S., COLE, F. N. and CUMMINGS, L. D., Complete classification of triad systems on fifteen elements, *Memoirs of National Academy of Sciences* **14,** Second Memoir, 1925.

[194] WINDLEY, P. F., Transposing matrices in a digital computer, *Comput. J.* **2,** 47 (1959).

[195] WUNDERLICH, M. C., Sieve-generated sequences, *Canad. J. Math.* **18,** 291 (1966).

[196] WUNDERLICH, M. C., Sieving procedures on a digital computer, *J. ACM* **14,** 10 (1967).

CHAPTER 7

[197] ALANEN, J., ORE, O. and STEMPLE, J., Systematic computations on amicable numbers, *Math. Comp* **21,** 242 (1967).

[198] BERLEKAMP, E. R., Factoring polynomials over finite fields, *Bell System Tech. J.* **46,** 1853 (1967).

[199] BERNHART, A., Six-rings in minimal five-color maps, *Amer. J. Math.* **69,** 391 (1947).

[200] BERNSTEIN, H. J., Machine factorization of groups, *Math. Algorithms* **1,** 39 (1966).

[201] BIRKHOFF, G. D., The reducibility of maps, *Amer. J. Math.* **35,** 115 (1913).

[202] BOSE, R. C., CHAKRAVARTI, I. M. and KNUTH, D. E., On methods of constructing sets of mutually orthogonal Latin squares using a computer, I, *Technometrics* **2,** 507 (1960).

[203] BOSE, R. C., SHRIKHANDE, S. S. and PARKER, E. T., Further results on the construction of mutually orthogonal Latin squares and the falsity of Euler's conjecture, *Canad. J. Math.* **12,** 189 (1960).

[204] BRILLHART, J., LEHMER, D. H. and LEHMER, E., Bounds for pairs of consecutive seventh to higher power residues, *Math. Comp.* **18,** 397 (1964).

[205] BROWN, J. W., Enumeration of Latin squares with application to order 8, *J. Combinatorial Theory* **5,** 177 (1968).

[206] CSIMA, J. and GOTLIEB, C. C., Tests on a computer method for constructing school timetables, *Comm. ACM* **7,** 160 (1964).

[207] DIXON, J. D., High speed computation of group characters, *Numer. Math.* **10,** 446 (1967).

[208] DULMAGE, A. L. and MENDELSOHN, N. S., Remarks on solutions of the optimal assignment problem, *SIAM J. Appl. Math.* **11,** 1103 (1963).

[209] EDMONDS, J., Covers and packings in a family of sets, *Bull. Amer. Math. Soc.* **68,** 494 (1962).

[210] EVANS, J. W., HARARY, F. and LYNN, M. L., On the computer enumeration of finite topologies, *Comm. ACM* **10,** 295 (1967).

[211] FLETCHER, J. G., A program to solve the pentomino problem by the recursive use of macros, *Comm. ACM* **8,** 621 (1965). Also see GOLOMB, S. W., References to pentominoes (letter to the editor), *Comm. ACM* **9,** 241 (1966).

[212] FORD, L. R., JR. and JOHNSON, S. M., A tournament problem, *Amer. Math. Monthly* **66,** 387 (1959).

[213] FRANKLIN, P., Note on the four color problem, *J. Math. Phys.* **16,** 172 (1938).

[214] GERHARDS, L. and LINDENBERG, W., A method for calculating the complete subgroup structure of finite groups using a binary computer, *Numer. Math.* **7,** 1 (1965).

[215] GOTLIEB, C. C. and CORNEIL, D. G., Algorithms for finding a fundamental set of cycles for an undirected linear graph, *Comm. ACM* **12,** 780 (1967).

[216] GRAVER, J. E. and YACKEL, J., Some graph theoretic results associated with Ramsey's theorem, *J. Combinatorial Theory* **4,** 125 (1968).

[217] GROSS, W. and VACCA, R., Distribution of the figures 0 and 1 in the various orders of binary representations of kth powers of integers, *Math. Comp.* **22,** 423 (1968).

[218] HAKIMI, S. L., On the degrees of vertices of a directed graph, *J. Franklin Inst.* **279,** 290 (1965).

[219] HALL, M., JR. and SENIOR, J. K., *The groups of order 2^n ($n \leqslant 6$)*. Macmillan, New York, 1964.

[220] HALL, M., JR., SWIFT, J. D. and KILLGROVE, R., On projective planes of order nine, *Math. Comp.* **13,** 233 (1959).

[221] HAYASHI, H. S., Computer investigation of difference sets, *Math. Comp.* **19,** 73 (1965).

[222] HEAWOOD, P. J., Map-colour theorem, *Quarterly J. of Math.* **24,** 332 (1890).

[223] IVANESCU, P. L. and ROSENBERG, I., Application of pseudo-Boolean programming to the theory of graphs, *Z. Wahrscheinlichkeitstheorie* **3**, 163 (1964).

[224] KEEDWELL, A. D., A search for projective planes of a special type with the aid of a digital computer, *Math. Comp.* **19**, 317 (1965).

[225] KEMPE, A. B., On the geographical problem of the four colours, *Amer. J. Math.* **2**, 193 (1879).

[226] KURTZBERG, J. M., On approximation methods for the assignment problem, *J. Assoc. Comput. Mach.* **9**, 419 (1962).

[227] LAWLER, E. L., An approach to multilevel Boolean minimization, *J. Assoc. Comput. Mach.* **11**, 283 (1964).

[228] LEHMER, D. H., Automation and pure mathematics, page 219 in reference [15].

[229] LEHMER, E., Number theory on the Swac, page 103 in reference [13].

[230] MANACHER, G. K., A bench mark calculation for retrieval by confluence of binary attributes in a random-access computer, *U. of Chicago Institute for Computer Research Quarterly Report* **14**, II A-1 (1967).

[231] MAURER, W. D., Computer experiments in finite algebra, *Comm. ACM* **9**, 598 (1966).

[232] MAYOH, B. H., On the simplification of logical expressions, *SIAM J. Appl. Math.* **15**, 898 (1967).

[233] NEUBAUER, G., An empirical investigation of Merten's function, *Numer. Math.* **5**, 1 (1963).

[234] NIKOLAI, P. J., Permanents of incidence matrices, *Math. Comp.* **14**, 262 (1960).

[235] ORE, O., *The Four-color Problem*, Academic, New York, 1967.

[236] PARKER, E. T., Computer investigation of orthogonal Latin squares of order ten, page 73 in reference [31].

[237] PARKER, E .T., Orthogonal Latin squares, *Proc. Nat. Acad. Sci. U.S.A.* **45**, 859 (1959).

[238] PARKER, E. T. and NIKOLAI, P. J., A search for analogues of the Mathieu groups, *Math. Tables Aids Comput.* **12**, 38 (1958).

[239] PARKIN, T. R., LANDER, L. J. and PARKIN, D. R., *Polyomino Enumeration Results*, presentation at SIAM Fall Meeting, December 1967, Santa Barbara, Calif.

[240] PLEMMONS, R. J., On computing non-equivalent finite algebraic systems, *Math. Algorithms* **2**, 80 (1967).

[241] RAY-CHADHURI, D. K., An algorithm for a minimum cover of an abstract complex, *Canad. J. Math.* **15**, 11 (1963).

[242] ROTH, J. P. and WAGNER, E. G., Algebraic topological methods for the synthesis of switching systems, III—Minimization of nonsingular Boolean trees, *IBM J. Res. Develop.* **3**, 326 (1959).

[243] SADE, A., *Énumération des Carrés Latins. Application au 7ᵉ Ordre*, Marseille, 1948.

[244] SCHURMANN, A., The application of graphs to the analysis of distribution of loops in a program, *Inf. Contr.* **7**, 275 (1964).

[245] SHAPIRO, M. B., An algorithm for reconstructing protein and RNA sequences, *J. Assoc. Comput. Mach.* **14**, 720 (1967).

[246] SILVER, R., The group of automorphisms of the game of 3-dimensional ticktacktoe, *Amer. Math. Monthly* **74**, 247 (1967).

[247] SLEPIAN, D., On the number of symmetry types of Boolean functions of n variables, *Canad. J. Math.* **5**, 185 (1953).

[248] STEINHAUS, H., *Mathematical Snapshots*, Oxford U.P., New York, 1960.

[249] SUSSENGUTH, E. H., JR., A graph-theoretic algorithm for matching chemical structures, *J. Chem. Doc.* **5**, 36 (1965).

[250] UNGER, S. H., GIT-heuristic program for testing pairs of directed line graphs for isomorphism, *Comm. ACM* **7**, 26 (1964).

[251] WALKER, R. J., Determination of division algebras with 32 elements, page 83 in reference [31].

[252] WATSON, E. J., Primitive polynomials (mod 2), *Math. Comp.* **16**, 368 (1962).

[253] WELLS, M. B., Application of a finite set covering theorem to the simplification of Boolean function expressions, page 731 in *Proc. of IFIP Congress 62*, C. M. Popplewell, editor, North-Holland, Amsterdam, 1963.

[254] WELLS, M. B., The number of Latin squares of order eight, *J. Combinatorial Theory* **3**, 98 (1967).

[255] WILLIAMSON, J., Note on Hadamard's determinant theorem, *Bull. Amer. Math. Soc.* **53**, 608 (1947).

[256] WINN, C. E., On certain reductions in the four-color problem, *J. Math. Phys.* **16**, 159 (1938).

INDEX